Application of Uncertainty Analysis to Ecological Risks of Pesticides

Other Titles from the Society of Environmental Toxicology and Chemistry (SETAC)

Linking Aquatic Exposure and Effects: Risk Assessment of Pesticides
Brock, Alix, Brown, Capri, Gottesbüren, Heimbach, Lythgo, Schulz, Streloke, editors
2009

Derivation and Use of Environmental Quality and Human Health Standards for Chemical Substances in Water and Soil
Crane, Matthiessen, Maycock, Merrington, Whitehouse, editors
2009

Aquatic Macrophyte Risk Assessment for Pesticides
Maltby, Arnold, Arts, Davies, Heimbach, Pickl, Poulsen, editors
2009

Veterinary Medicines in the Environment
Crane, Boxall, Barrett
2008

Relevance of Ambient Water Quality Criteria for Ephemeral and Effluent-dependent Watercourses of the Arid Western United States
Gensemer, Meyerhof, Ramage, Curley
2008

Extrapolation Practice for Ecotoxicological Effect Characterization of Chemicals
Solomon, Brock, de Zwart, Dyev, Posthumm, Richards, editors
2008

Environmental Life Cycle Costing
Hunkeler, Lichtenvort, Rebitzer, editors
2008

Valuation of Ecological Resources: Integration of Ecology and Socioeconomics in Environmental Decision Making
Stahl, Kapustka, Munns, Bruins, editors
2007

For information about SETAC publications, including SETAC's international journals, Environmental Toxicology and Chemistry and Integrated Environmental Assessment and Management, contact the SETAC office nearest you:

SETAC
1010 North 12th Avenue
Pensacola, FL 32501-3367 USA
T 850 469 1500 F 850 469 9778
E setac@setac.org

SETAC Office
Avenue de la Toison d'Or 67
B-1060 Brussells, Belguim
T 32 2 772 72 81 F 32 2 770 53 86
E setac@setaceu.org

www.setac.org
Environmental Quality Through Science®

Application of Uncertainty Analysis to Ecological Risks of Pesticides

Edited by
William J. Warren-Hicks
Andy Hart

Coordinating Editor of SETAC Books
Joseph W. Gorsuch
Copper Development Association, Inc.
New York, NY, USA

CRC Press
Taylor & Francis Group
Boca Raton London New York

CRC Press is an imprint of the
Taylor & Francis Group, an **informa** business

Published in collaboration with the Society of Environmental Toxicology and Chemistry (SETAC)
1010 North 12th Avenue, Pensacola, Florida 32501
Telephone: (850) 469-1500 ; Fax: (850) 469-9778; Email: setac@setac.org
Web site: www.setac.org

CRC Press
Taylor & Francis Group
6000 Broken Sound Parkway NW, Suite 300
Boca Raton, FL 33487-2742

First issued in paperback 2017

© 2010 by Taylor & Francis Group, LLC
CRC Press is an imprint of Taylor & Francis Group, an Informa business

No claim to original U.S. Government works

ISBN-13: 978-1-4398-0734-7 (hbk)
ISBN-13: 978-1-138-11481-4 (pbk)

Library of Congress Cataloging-in-Publication Data

Application of uncertainty analysis to ecological risks of pesticides / editors, William J. Warren-Hicks, Andy Hart.
    p. cm.
Includes bibliographical references and index.
ISBN 978-1-4398-0734-7 (hardcover : alk. paper)
    1. Pesticides--Environmental aspects--Mathematical models. 2. Ecological risk assessment. 3. Probabilities. I. Warren-Hicks, William J. II. Hart, Andy, 1956-

QH545.P4.A67 2010
577.27'9015118--dc22   2009038019

Visit the Taylor & Francis Web site at
http://www.taylorandfrancis.com

the CRC Press Web site at
http://www.crcpress.com

and the SETAC Web site at
www.setac.org

SETAC Publications

Books published by the Society of Environmental Toxicology and Chemistry (SETAC) provide in-depth reviews and critical appraisals on scientific subjects relevant to understanding the impacts of chemicals and technology on the environment. The books explore topics reviewed and recommended by the Publications Advisory Council and approved by the SETAC North America, Latin America, or Asia/Pacific Board of Directors; the SETAC Europe Council; or the SETAC World Council for their importance, timeliness, and contribution to multidisciplinary approaches to solving environmental problems. The diversity and breadth of subjects covered in the series reflect the wide range of disciplines encompassed by environmental toxicology, environmental chemistry, and hazard and risk assessment, and life-cycle assessment. SETAC books attempt to present the reader with authoritative coverage of the literature, as well as paradigms, methodologies, and controversies; research needs; and new developments specific to the featured topics. The books are generally peer reviewed for SETAC by acknowledged experts.

SETAC publications, which include Technical Issue Papers (TIPs), workshops summaries, newsletter (SETAC Globe), and journals (*Environmental Toxicology and Chemistry* and *Integrated Environmental Assessment and Management*), are useful to environmental scientists in research, research management, chemical manufacturing and regulation, risk assessment, and education, as well as to students considering or preparing for careers in these areas. The publications provide information for keeping abreast of recent developments in familiar subject areas and for rapid introduction to principles and approaches in new subject areas.

SETAC recognizes and thanks the past coordinating editors of SETAC books:

A.S. Green, International Zinc Association
Durham, North Carolina, USA

C.G. Ingersoll, Columbia Environmental Research Center
US Geological Survey, Columbia, Missouri, USA

T.W. La Point, Institute of Applied Sciences
University of North Texas, Denton, Texas, USA

B.T. Walton, US Environmental Protection Agency
Research Triangle Park, North Carolina, USA

C.H. Ward, Department of Environmental Sciences and Engineering
Rice University, Houston, Texas, USA

Contents

List of Figures

List of Tables

Foreword

The workshop from which this book resulted, Application of Uncertainty Analysis to Ecological Risks of Pesticides, was part of the successful Pellston Workshop Series. Since 1977, Pellston Workshops have brought scientists together to evaluate current and prospective environmental issues. Each workshop has focused on a relevant environmental topic, and the proceedings of each have been published as peer-reviewed or informal reports. These documents have been widely distributed and are valued by environmental scientists, engineers, regulators, and managers for their technical basis and their comprehensive, state-of-the-science reviews. The other workshops in the Pellston series are as follows:

- Estimating the Hazard of Chemical Substances to Aquatic Life. Pellston, Michigan, 13 to 17 Jun 1977. Published by the American Society for Testing and Materials, STP 657, 1978.
- Analyzing the Hazard Evaluation Process. Waterville Valley, New Hampshire, 14 to 18 Aug 1978. Published by The American Fisheries Society, 1979.
- Biotransformation and Fate of Chemicals in the Aquatic Environment. Pellston, Michigan, 14 to 18 Aug 1979. Published by The American Society of Microbiology, 1980.
- Modeling the Fate of Chemicals in the Aquatic Environment. Pellston, Michigan, 16 to 21 Aug 1981. Published by Ann Arbor Science, 1982.
- Environmental Hazard Assessment of Effluents. Cody, Wyoming, 23 to 27 Aug 1982. Published as a SETAC Special Publication by Pergamon Press, 1985.
- Fate and Effects of Sediment-Bound in Aquatic Systems. Florissant, Colorado, 11 to 18 Aug 1984. Published as a SETAC Special Publication by Pergamon Press, 1987.
- Research Priorities in Environmental Risk Assessment. Held in Breckenridge, Colorado, Aug 16 to 21, 1987. Published by SETAC, 1987.
- Biomarkers: Biochemical, Physiological, and Histological Markers of Anthropogenic Stress. Keystone, Colorado, 23 to 28 Jul 1989. Published as a SETAC Special Publication by Lewis Publishers, 1992.
- Population Ecology and Wildlife Toxicology of Agricultural Pesticide Use: A Modeling Initiative for Avian Species. Kiawah Island, South Carolina, 22 to 27 Jul 1990. Published as a SETAC Special Publication by Lewis Publishers, 1994.
- A Technical Framework for [Product] Life-Cycle Assessments. Smuggler's Notch, Vermont, 18 to 23 August 1990. Published by SETAC, Jan 1991; 2nd printing Sep 1991; 3rd printing Mar 1994.
- Aquatic Microcosms for Ecological Assessment of Pesticides. Wintergreen, Virginia, 7 to 11 Oct 1991. Published by SETAC, 1992.

- A Conceptual Framework for Life-Cycle Assessment Impact Assessment. Sandestin, Florida, 1 to 6 Feb 1992. Published by SETAC, 1993.
- A Mechanistic Understanding of Bioavailability: Physical–Chemical Interactions. Pellston, Michigan, 17 to 22 Aug 1992. Published as a SETAC Special Publication by Lewis Publishers, 1994.
- Life-Cycle Assessment Data Quality Workshop. Wintergreen, Virginia, 4 to 9 Oct 1992. Published by SETAC, 1994.
- Avian Radio Telemetry in Support of Pesticide Field Studies. Pacific Grove, California, 5 to 8 Jan 1993. Published by SETAC, 1998.
- Sustainability-Based Environmental Management. Pellston, Michigan, 25 to 31 Aug 1993. Co-sponsored by the Ecological Society of America. Published by SETAC, 1998.
- Ecotoxicological Risk Assessment for Chlorinated Organic Chemicals. Alliston, Ontario, Canada, 25 to 29 Jul 1994. Published by SETAC, 1998.
- Application of Life-Cycle Assessment to Public Policy. Wintergreen, Virginia, 14 to 19 Aug 1994. Published by SETAC, 1997.
- Ecological Risk Assessment Decision Support System. Pellston, Michigan, 23 to 28 Aug 1994. Published by SETAC, 1998.
- Avian Toxicity Testing. Pensacola, Florida, 4 to 7 Dec 1994. Co-sponsored by Organisation for Economic Co-operation and Development. Published by OECD, 1996.
- Chemical Ranking and Scoring (CRS): Guidelines for Developing and Implementing Tools for Relative Chemical Assessments. Sandestin, Florida, 12 to 16 Feb 1995. Published by SETAC, 1997.
- Ecological Risk Assessment of Contaminated Sediments. Pacific Grove, California, 23 to 28 Apr 1995. Published by SETAC, 1997.
- Ecotoxicology and Risk Assessment for Wetlands. Fairmont, Montana, 30 Jul to 3 Aug 1995. Published by SETAC, 1999.
- Uncertainty in Ecological Risk Assessment. Pellston, Michigan, 23 to 28 Aug 1995. Published by SETAC, 1998.
- Whole-Effluent Toxicity Testing: An Evaluation of Methods and Prediction of Receiving System Impacts. Pellston, Michigan, 16 to 21 Sep 1995. Published by SETAC, 1996.
- Reproductive and Developmental Effects of Contaminants in Oviparous Vertebrates. Fairmont, Montana, 13 to 18 Jul 1997. Published by SETAC, 1999.
- Multiple Stressors in Ecological Risk Assessment. Pellston, Michigan, 13 to 18 Sep 1997. Published by SETAC, 1999.
- Re-Evaluation of the State of the Science for Water Quality Criteria Development. Fairmont, Montana, 25 to 30 Jun 1998. Published by SETAC, 2003.
- Criteria for Persistence and Long-Range Transport of Chemicals in the Environment. Fairmont Hot Springs, British Columbia, Canada, 14 to 19 Jul 1998. Published by SETAC, 2000.

- Assessing Contaminated Soils: From Soil-Contaminant Interactions to Ecosystem Management. Pellston, Michigan, 23 to 27 Sep 1998. Published by SETAC, 2003.
- Endocrine Disruption in Invertebrates: Endocrinology, Testing, and Assessment (EDIETA). Amsterdam, The Netherlands, 12 to 15 Dec 1998. Published by SETAC, 1999.
- Assessing the Effects of Complex Stressors in Ecosystems. Pellston, Michigan, 11 to 16 Sep 1999. Published by SETAC, 2001.
- Environmental–Human Health Interconnections. Snowbird, Utah, 10 to 15 Jun 2000. Published by SETAC, 2002.
- Ecological Assessment of Aquatic Resources: Application, Implementation, and Communication. Pellston, Michigan, 16 to 21 Sep 2000. Published by SETAC, 2004.
- Toxicity Identification Evaluation/Toxicity Reduction Evaluation: What Works and What Doesn't. Pensacola, Florida, 23 to 27 Jun 2001. Published by SETAC, 2005.
- The Global Decline of Amphibian Populations: An Integrated Analysis of Multiple Stressors Effects. Wingspread, Racine, Wisconsin, 18 to 23 Aug 2001. Published by SETAC, 2003.
- The Role of Dietary Exposure in the Evaluation of Risk of Metals to Aquatic Organisms. Fairmont Hot Springs, British Columbia, Canada, 27 Jul to 1 Aug 2002. Published by SETAC, 2005.
- Use of Sediment Quality Guidelines (SQGs) and Related Tools for the Assessment of Contaminated Sediments. Fairmont Hot Springs, Montana, 17 to 22 Aug 2002. Published by SETAC, 2005.
- Science for Assessment of the Impacts of Human Pharmaceuticals on Aquatic Ecosystem. Snowbird, Utah, 3 to 8 Jun 2003. Published by SETAC, 2005.
- Population-Level Ecological Risk Assessment. Roskilde, Denmark, 23 to 27 Aug 2003. Published by SETAC and CRC Press, 2007.
- Valuation of Ecological Resources: Integration of Ecological Risk Assessment and Socio-Economics to Support Environmental Decisions. Pensacola, Florida, 4 to 9 Oct 2003. Published by SETAC and CRC Press, 2007.
- Emerging Molecular and Computational Approaches for Cross-Species Extrapolations. Portland, Oregon, 18 to 22 Jul 2004. Published by SETAC and CRC Press, 2006.
- Molecular Biology and Risk Assessment: Evaluation of the Potential Roles of Genomics in Regulatory Ecotoxicology. Pellston, Michigan, 18 to 22 Sep 2006. Published by SETAC and CRC Press, 2008.
- Veterinary Medicines in the Environment. Pensacola, Florida, 12 to 16 Feb 2006. Published by SETAC and CRC Press, 2008.
- Tissue Residue Approach for Toxicity Assessment: Invertebrates and Fish. Leavenworth, Washington, 7 to 10 Jun 2007. To be published by SETAC and CRC Press, 2010.
- Science-Based Guidance and Framework for the Evaluation and Identification of PBTs and POPs. Pensacola Beach, Florida, 27 Jan to 1

Feb 2008. To be published in the SETAC journal *Integrated Environmental Assessment and Management* (IEAM) in 2009.

- Ecological Assessment of Selenium in the Aquatic Environment. Pensacola, Florida, 23 to 28 Feb 2009. To be published by SETAC and CRC Press, 2010.

Acknowledgments

We gratefully acknowledge the financial support for the workshop by the following organizations:

American Chemistry Council
American Crop Protection Association
Bayer CropScience
Dow AgroSciences LLC
DuPont
Health Canada
Syngenta
US Environmental Protection Agency

About the Editors

William J. Warren-Hicks is CEO of EcoStat, Inc., a small company he cofounded in 2004. He holds a PhD from Duke University in environmental statistics. He has 29 years of experience providing consulting expertise in the areas of environmental data analysis, uncertainty analysis, Bayesian inference and decision, probabilistic risk methods, survey design, time-series modeling, messy data analysis, hypothesis testing, multivariate analyses, and model validation studies. Dr. Warren-Hicks has participated in 5 SETAC Pellston Workshops, including Sediment Risk Assessment (chapter author), Multiple Stressors (author and steering committee member), Probabilistic Risk Assessment of Pesticides (book editor and author), Whole Effluent Toxicity Testing (chapter author), and Uncertainty Analysis in Ecological Risk Assessment (chair, lead editor, lead conference organizer, and author). As a SETAC member since 1988, he has presented numerous papers, organized and presented numerous short courses, and organized and chaired numerous sessions at the SETAC annual meetings. Dr. Warren-Hicks is credited with more than 120 peer-reviewed publications addressing the application of probability and statistics in water, air, and terrestrial systems.

Andy Hart is at the Food and Environment Research Agency (FERA), an agency of the UK Department for Environment, Food and Rural Affairs. He leads FERA's Risk Analysis Team, which focuses on developing and applying improved approaches for dealing with variability and uncertainty in human and environmental risk assessment. Dr. Hart was a member of the USEPA's ECOFRAM committee, developing probabilistic approaches for assessment of pesticide risks to birds, and played a leading role in the UK WEBFRAM projects, which developed online probabilistic models for pesticide risk assessment. He initiated and chaired the SETAC Pellston Workshop on uncertainty analysis in pesticide risk assessment, which produced this book. Together with his team, he is applying probabilistic approaches in other areas, including human exposure to food contaminants, invasive species, and risk–benefit analysis. Dr. Hart is a member of the European Food Safety Authority's expert panel on pesticides (EFSA PPR Panel) and was also a member of EFSA and WHO working groups that developed guidance documents on uncertainty analysis for human dietary exposure assessment.

Workshop Participants* and Contributing Authors†

Tom Aldenberg*†
RIVM
Bilthoven, The Netherlands

Timothy Barry*†
USEPA
Washington DC, USA

Theo C. M. Brock*†
Alterra Green World Research
Wageningen, The Netherlands

Bryan W. Brooks*†
Baylor University
Waco, Texas, USA

Lawrence A. Burns*†
USEPA
Athens, Georgia, USA

John P. Carbone*†
Rohm & Haas Company
Spring House, Pennsylvania, USA

Peter F. Chapman*†
Syngenta Ecological Science
Bracknell, Berkshire, UK

Karin Corsten*†
Federal Office of Consumer Protection
    and Food Safety (BVL)
Braunschweig, Germany

Mark Crane*†
wca environment limited
Faringdon, Oxfordshire, UK

Peter L. deFur*†
Environmental Stewardship Concepts
Richmond, Virginia, USA

Peter D. Delorme*†
PMRA, Health Canada
Ottawa, Ontario, Canada

Tammara L. Estes*†
Stone Environmental, Inc.
Wilmette, Illinois, USA

David A. Evans†
College of William and Mary
Gloucester Point, Virginia, USA

Anne Fairbrother*†
Exponent, Inc.
Bellevue, Oregon, USA

David Farrar*†
USEPA
Washington DC, USA

Scott Ferson*†
Applied Biomathematics
Setauket, New York, USA

David L. Fischer*†
Bayer CropScience
Research Triangle Park, North
    Carolina, USA

Edward Fite*†
USEPA
Washington DC, USA

* Affiliations have been updated, where possible, since the time of the workshop.

Kathryn Gallagher*†
USEPA
Washington DC, USA

Jeffrey M. Giddings*†
Compliance Services International
Rochester, Massachusetts, USA

Mick Hamer*†
Syngenta
Bracknell, Berkshire, UK

Andrew Hart*†
Food and Environment Research
    Agency
Sand Hutton, York, UK

Wolfgang Heger*†
Umweltbundesamt
Berlin, Germany

Paul Hendley*†
Syngenta Crop Protection, Inc.
Greensboro, North Carolina, USA

Joanna Jaworska*†
Procter & Gamble
Strombeek-Bever, Belgium

Gerhard Joermann*†
Federal Office of Consumer Protection
    and Food Safety (BVL)
Braunschweig, Germany

Paul D. Jones*†
University of Saskatchewan
Saskatoon, Saskatchewan, Canada

Denise M. Keehner*†
USEPA
Washington DC, USA

Charalyn Kriz*†
Health Canada
Ottawa, Ontario, Canada

Thomas W. La Point*†
University of North Texas
Denton, Texas, USA

Wayne G. Landis*†
Western Washington University
Bellingham, Washington, USA

Michael Lavine*†
University of Massachusetts
Amherst, Massachusetts, USA

Robert Luttik*†
RIVM
Bilthoven, The Netherlands

Pierre Mineau*†
Canadian Wildlife Service
Ottawa, Ontario, Canada

Dwayne R. J. Moore*†
Intrinsik Environmental Services
Ottawa, Ontario, Canada

Michael C. Newman*†
Virginia Institute of Marine Sciences
Glouster Point, Virginia, USA

Raymond J. O'Connor*†
University of Maine
Orono, Maine, USA

Edward W. Odenkirchen*†
USEPA
Washington DC, USA

Song S. Qian†
Duke University
Durham, North Carolina, USA

Hans-Toni Ratte*†
RWTH Aachen University
Aachen, Germany

Kees Romijn†
Bayer CropScience
Monheim, Germany

Mark H. Russell*†
DuPont Crop Protection
Newark, Delaware, USA

Jennifer L. Shaw*†
Syngenta Crop Protection, Inc.
Greensboro, North Carolina, USA

Eric P. Smith*†
Virginia Tech
Blacksburg, Virginia, USA

Glenn W. Suter*†
USEPA
Cincinnati, Ohio, USA

John Toll*†
Worldward Environmental, LLC
Seattle, Washington, USA

Theo P. Traas*†
RIVM
Bilthoven, The Netherlands

Kelley R. Tucker*†
American Bird Conservancy
Washington, DC, USA

Doug Urban†
USEPA
Washington DC, USA

Paul J. Van den Brink*†
Alterra and Wageningen University
Wageningen, The Netherlands

Frederik Verdonck*†
Ghent University
Ghent, Belgium

William J. Warren-Hicks*†
EcoStat, Inc.
Mebanc, North Carolina, USA

1 Introduction and Objectives

A. Hart, D. Farrar, D. Urban, D. Fischer,
T. La Point, K. Romijn, and S. Ferson

1.1 INTRODUCTION

Current methods used in ecological risk assessments for pesticides are largely "deterministic." These methods generally produce simple measures of risk (e.g., risk quotients) and do not quantify the influence of variability and uncertainty in exposure and effects. "Probabilistic" methods do quantify and analyze variability and uncertainty. They can also provide more meaningful measures of risk (e.g., the frequency and magnitude of impacts). Consequently, probabilistic methods are attracting growing interest from both industry and government, especially in North America (USEPA 2000), but also in Europe (e.g., Hart 2001), and elsewhere.

Uncertainty analysis is increasingly used in ecological risk assessment and was the subject of an earlier Pellston workshop (Warren-Hicks and Moore 1998). The US Environmental Protection Agency (USEPA) has developed general principles for the use of Monte Carlo methods (USEPA 1997), which provide one of several approaches to incorporating variability and uncertainty in risk assessment.

This chapter considers the role of variability and uncertainty in ecological risk assessment and discusses whether it is necessary to quantify them. It concludes by setting out the objectives and key issues that were considered at the Pellston workshop in February 2002, which are addressed in the following chapters of this book.

1.2 VARIABILITY AND UNCERTAINTY

A variety of terms are used in the statistical literature to define and describe the general concept of imperfect knowledge. Many authors prefer specific terms that are explicitly defined, while others use more general terminology that can be applied in a broad context. Two terms that are of general use are (Suter and Barnthouse 1993):

> *Stochasticity:* the inherent randomness of the world,
> *Ignorance:* imperfect or incomplete knowledge of things that could be known.

There are many ways of classifying the various types of uncertainty and variability that are associated with these two terms. In the peer-reviewed literature, it is

TABLE 1.1

Authors' terms for stochasticity, tandomness, ignorance, or doubt

Reference	Stochasticity or randomness	Ignorance or doubt	Combination of both
Hattis and Burmaster (1994)	Variability	Uncertainty	Variability and uncertainty
Apostolakis (1994, 1999)	Aleatory uncertainty	Epistemic uncertainty	Uncertainty
Ferson and Ginzburg (1996)	Variability	Incertitude	Uncertainty
Hoffman and Hammonds (1994)	Type A uncertainty	Type B uncertainty	Uncertainty
Klir and Yuan (1995)	Conflict	Nonspecificity	Ambiguity
Walley (1991)	Chance	Imprecision	Imprecise probability

not uncommon to find different authors and practitioners using various terms for describing stochasticity and ignorance. Table 1.1 indicates some approaches.

Norton (1998) offers one approach from an ecological risk assessor's perspective. Various experts in statistics and risk assessment at the workshop preferred specific terms over others. The majority of the workshop participants were comfortable with distinguishing between uncertainty and variability in a manner that is consistent with US Environmental Protection Agency (USEPA) guidance. Other experts, particularly those associated with bounding analyses (see Chapter 6 of this book), preferred the word incertitude instead of uncertainty based on theoretical considerations associated with the bounding methods discussed in the chapter.

In order to simplify the terminology in this book, most chapters will use the term uncertainty to refer to doubt or ignorance, due to imperfect or incomplete knowledge of things that could be known; for example, uncertainty about the true mean of a population when only a small sample of the population has been measured, or incomplete understanding of the mode of action of a toxicant. And, all chapters will use the term variability to refer to stochasticity and heterogeneity; the existence of natural and anthropogenic variation in the real world, including differences between individuals, spatial variation, and changes over time.

Readers of this book are encouraged to consult the references shown in Table 1.1 to obtain additional information on the concepts of variability, uncertainty, incertitude, imprecision, chance, ambiguity, and other terms that arise in ecological uncertainty analysis.

1.3   IMPORTANCE OF VARIABILITY AND UNCERTAINTY IN RISK ASSESSMENT

Variability and uncertainty affect every element of every risk assessment. For example, participants in the European Workshop on Probabilistic Risk Assessment for the Environmental Impacts of Plant Protection Products (EUPRA) were asked to list sources of uncertainty affecting current procedures for assessing pesticide risks to aquatic organisms, terrestrial vertebrates, terrestrial invertebrates, and plants (Hart 2001). The resulting lists include many diverse sources of uncertainty and variability in both exposure assessment and effects assessment for all 3 groups of organisms (Table 1.2).

Many sources of stochasticity and uncertainty are large enough to change the level of risk by orders of magnitude. For example, pesticide residues on terrestrial invertebrates after spraying vary over 2 orders of magnitude (Figure 1.1). If such important sources of variation were ignored, the resulting assessment would give a very misleading picture of the true range of risks.

TABLE 1.2

Some key sources of uncertainty affecting current risk assessments for pesticides in Europe, as listed by the EUPRA workshop (Hart 2001)

Aquatic organisms

Extrapolation from individuals in single species tests to populations

Extrapolation from individuals of single species to communities

Input parameters for modeling pesticide fate

Discrepancy between exposure in laboratory studies and in the field

Uncertainties in the exposure scenario and variability in the landscape

Variation in sensitivity between species

Extrapolation of sensitivity from laboratory studies to the field

Representativeness of species used in risk assessment

Level of effect that is acceptable

Influence of indirect effects

Terrestrial vertebrates

Intraspecies and interspecies variation in sensitivity

Behavior and natural history

Spatial distribution of residues

Residues dynamics (dissipation, bioaccumulation, etc.)

Avoidance or attraction of contaminated food

Effects on populations and communities

Mismatch between exposure pattern over time in laboratory and field

Nondietary routes of exposure (e.g., dermal exposure and inhalation)

Terrestrial invertebrates and plants

Level of effect that is acceptable

Factors affecting exposure

Interspecies variation in sensitivity

Extrapolation from effects on individuals (lab) to populations (field)

Extrapolating acute to chronic effects

Regional variation in sensitivity or concern for nontarget organisms

Errors in the structure of risk assessment models

Presence of sensitive life stages

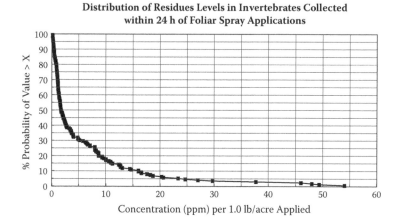

FIGURE 1.1   Variation in residues of pesticides on terrestrial invertebrates (D. Fischer, pers. comm.).

Similarly, uncertainty can also have substantial implications for assessment outcomes. For example, there is uncertainty about the importance of exposure of birds to pesticides via dermal absorption. This route of exposure is generally ignored in regulatory assessments, but there is evidence that it may be as important as dietary exposure, at least in some circumstances (Driver et al. 1991; Mineau 2002). This uncertainty, therefore, implies a potential 2-fold error in the assessment of exposure.

It is therefore important to take account of variability and uncertainty in risk assessment. The question is, how?

1.4   CURRENT METHODS FOR DEALING WITH VARIABILITY AND UNCERTAINTY ARE INADEQUATE

The most common methods for dealing with variability and uncertainty in the past have been the use of conservative assumptions, safety factors, and assessment scenarios. Each of these approaches has limitations that may often lead to inappropriate decisions.

1.4.1   CONSERVATIVE ASSUMPTIONS

Conservative or "worst-case" assumptions are very commonly used. However, the degree of conservatism varies between assumptions and is rarely quantified (Figure 1.2). Furthermore, when many assumptions are combined in the same assessment, the overall degree of conservatism is difficult to determine.

1.4.2   SAFETY FACTORS

Safety or uncertainty factors are often applied at the end of an assessment, for example, as a "level of concern" to which a risk quotient or toxicity-exposure ratio is compared.

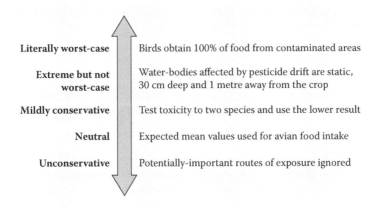

Literally worst-case	Birds obtain 100% of food from contaminated areas
Extreme but not worst-case	Water-bodies affected by pesticide drift are static, 30 cm deep and 1 metre away from the crop
Mildly conservative	Test toxicity to two species and use the lower result
Neutral	Expected mean values used for avian food intake
Unconservative	Potentially-important routes of exposure ignored

FIGURE 1.2 Examples of differing degrees of conservatism in assumptions used in pesticide risk assessments.

Often, the basis for safety factors is obscure and/or arbitrary. They are typically based on order of magnitude decisions, for example, determining a "safe" concentration and dividing it by 10. It may not be clear which sources of uncertainty they are intended to address. Even when they have been based on an explicit assessment of uncertainty, this probably will not have included more than a few sources of uncertainty. Therefore, it is not known whether the safety factors provide an appropriate level of protection against all the uncertainties affecting the assessment.

1.4.3 Assessment Scenarios

Sometimes, different assessment scenarios are used, for example, to assess risk under contrasting environmental conditions in different geographical areas, or to assess risk to different receptors (e.g., insectivorous vs. herbivorous species of birds). However, it is rarely practical to use more than a small number of scenarios and it is very difficult to determine how well the chosen scenarios represent the full range of true scenarios, especially if the scenarios differ with respect to many variables.

1.5 VARIABILITY AND UNCERTAINTY HINDER THE REGULATORY PROCESS

The inadequacy of current approaches for dealing with variability and uncertainty is currently causing significant practical difficulties in regulatory procedures, including

- Disputes among stakeholders about just how conservative the assessments are
- Difficulty in identifying what types of additional data are required to reduce uncertainty

- Lack of agreement about whether and how much to alter safety factors, when extra data are provided

These problems cause delays in regulatory decision making, which have significant implications for all the stakeholders.

1.6 UNDERSTANDING UNCERTAINTY AND VARIABILITY IS CRITICAL WHEN DEVELOPING A CREDIBLE RISK ASSESSMENT

Ineffective or unconvincing approaches to uncertainty can affect the credibility of individual risk assessments or the regulatory process as a whole.

This has been highlighted by a number of food safety issues. For example, lack of confidence that uncertainties were being adequately dealt with has been an important factor in recent public concerns about bovine spongiform encephalopathy and genetically modified crops, especially in Europe. Uncertainty was also a factor in the earlier controversy over alar in apples (e.g., Ames and Gold 1989; Groth 1989; Thayer 1989).

Uncertainty may also affect the credibility of ecological risk assessment procedures. In the late 1970s, the USEPA presented a risk assessment for the use of granular carbofuran on corn, including a detailed list of field studies and incidents. The Federal Insecticide, Fungicide, and Rodenticide Act Science Advisory Panel concluded there was insufficient information to justify restricted use labeling and recommended further testing. Nearly 20 years later, the accumulation of additional field studies and incidents provided sufficient evidence such that approvals for use of carbofuran were withdrawn.

1.7 QUANTITATIVE ANALYSIS OF VARIABILITY AND UNCERTAINTY CAN HELP

Quantitative analysis can help by

- Identifying and quantifying known sources of variability and uncertainty
- Showing the consequences of known sources of variability and uncertainty for the overall assessment
- Focusing attention on the most important known sources of variability and uncertainty

A previous Pellston workshop listed the benefits of uncertainty analysis in regulatory programs as follows (Warren-Hicks and Moore 1998):

- Improved transparency
- Improved credibility
- Better focusing of data collection
- Avoidance of worst-case assumptions
- Improved basis for decision making

1.8   WHEN IS QUANTITATIVE ANALYSIS OF VARIABILITY AND UNCERTAINTY REQUIRED?

The issues considered in the preceding sections imply that variability and uncertainty should be considered in some way in every risk assessment.

Screening assessments incorporate variability and uncertainty implicitly, by using worst-case assumptions and safety factors. As mentioned earlier, these have rarely been based on a quantitative analysis and may not take account of the full range of uncertainties, so in principle they should be reviewed to determine whether they provide adequate margins of safety.

In higher tier assessments, the question is not so much whether uncertainty analysis is required, but rather whether it should be quantitative and what methods should be used for it. The previous Pellston workshop made the following recommendations as a general guide (Warren-Hicks and Moore 1998):

- Quantitative uncertainty analysis is not appropriate when in a worst-case approach, risk is found to be negligible; when field evidence indicates obvious and severe effects; when information is insufficient to adequately characterize the model equation, input probability density functions (PDFs), and the relationships between the PDFs; or when it is more cost-effective to take action than to conduct more analyses.
- Quantitative uncertainty analysis is appropriate when it is essential to set priorities among sites, contaminants, exposure pathways, receptors, or other risk factors, given limited resources; the consequences of an incorrect decision are high; and available or obtainable information is insufficient to conduct a defensible analysis.

A variety of methods are available for analyzing variability and uncertainty quantitatively. Later chapters describe the main approaches and provide guidance on how to decide which is appropriate for a particular case.

1.9   WHAT IF THE BOUNDS ARE VERY WIDE?

A potential concern about quantifying uncertainty has been that it may generate such wide bounds on risk estimates as to make them unusable for decision making. This, together with the greater complexity of quantitative methods, has led to suggestions that it might be better to use simple hazard assessment with large safety factors. However, using simple safety factors will not produce narrower bounds on risk estimates unless the safety factors understate the true level of uncertainty. On the contrary, a specific quantitative analysis will often produce narrower bounds than generalized safety factors, because the latter should be sufficiently large to take appropriate account of all cases. When the bounds on risk are wide, decision makers can either ask for further research to reduce uncertainty or make a decision (either precautionary or otherwise, as appropriate) that takes account of the uncertainty.

1.10   NEED FOR CONSENSUS ON APPROPRIATE METHODS

There are many methods of analyzing variability and uncertainty and many ways of presenting the results. Inappropriate use of these methods gives misleading results, and experts differ on what is appropriate. Disagreement about which methods are appropriate will lead to wasted resources, conflict over results, and reduced credibility with decision makers and the public. There is, therefore, a need to reach a consensus on how to choose and use appropriate methods, and to present this in the form of guidance for prospective users.

1.11   WORKSHOP OBJECTIVES AND KEY ISSUES

The Pellston workshop in February 2002, which produced this book, aimed to develop guidance and increased consensus on the use of uncertainty analysis methods in ecological risk assessment. The workshop focused on pesticides, and used case studies on pesticides, because of the urgent need created by the rapid move to using probabilistic methods in pesticide risk assessment. However, it was anticipated that the conclusions would also be highly relevant to other stressors, especially other contaminants.

1.11.1   WORKSHOP OBJECTIVES

- Promote wider understanding of uncertainty analysis, especially in the pesticides arena, by providing an accessible review of the main approaches
- Provide guidance on how to select appropriate methods of uncertainty analysis, and how to use them
- Develop case studies to explore the application of alternative methods of uncertainty analysis to the ecological risks of pesticides
- Identify priorities for further development, implementation, and training

1.11.2   KEY ISSUES

In addressing its objectives and developing the case studies, the workshop gave particular consideration to the following key issues.

Which methods for analyzing variability and uncertainty are appropriate under what circumstances? Which methods are appropriate when data are limited? What are the strengths and weaknesses of different methods? What are their main principles and pitfalls?

- What are the implications of probabilistic methods for problem formulation?
- How can uncertainty analysis methods be used to help achieve the desired level of certainty efficiently?

- When and how should we separate variability and incertitude, or partition uncertainty in other ways?
- How can we take account of uncertainty concerning the structure of the conceptual model for the assessment?
- How should we select and parameterize input distributions when data are limited?
- How should we deal with dependencies, including nonlinear dependencies and dependencies about which only partial information is available?
- How can we take account of uncertainty when combining different types of information in an assessment (e.g., quantitative data and expert judgment, laboratory data and field data)?
- How can we detect and avoid misleading results?
- How can we communicate methods and outputs effectively to decision makers and stakeholders?
- What are the priorities for further development, implementation, and training?

1.12 REFERENCES

Ames B, Gold L. 1989. Pesticides, risk, and applesauce. Science, Letters 244:755–757.

Apostolakis GE. 1994. A commentary on model uncertainty. In: Mosleh A, Siu N, Smidts C, Lui C, editors. Proceedings of Workshop I in Advanced Topics in Risk and Reliability Analysis, Model Uncertainty: Its Characterization and Quantification.

Apostolakis GE. 1999. The distinction between aleatory and epistemic uncertainties is important: an example from the inclusion of aging effects into probabilistic safety assessment. Proceedings of the PSA'99, August 22 to 25, 1999. Washington (DC): American Nuclear Society.

Driver CJ, Ligotke MW, Van Voris P, McVeety BD, Greenspan BJ, Drown DB. 1991. Routes of uptake and their relative contribution to the toxicological response of northern bobwhite (*Colinus virginianus*) to an organophosphate pesticide. Environ Toxicol Chem 10:21–33.

Ferson S, Ginzburg LR. 1996. Different methods are needed to propagate ignorance and variability. Reliability Eng Syst Saf 54:133–144.

Groth E. 1989. Alar in apples. Science, Letters 244:755.

Hart A. 2001. Probabilistic risk assessment for pesticides in Europe: implementation and research needs. In: Report of the European Workshop on Probabilistic Risk Assessment for the Environmental Impacts of Plant Protection Products (EUPRA). Central Science Laboratory, Sand Hutton, United Kingdom. 109 p. Available from: http://www.eupra.com.

Hattis D, Burmaster DE. 1994. Assessment of variability and uncertainty distributions for practical risk analyses. Risk Anal 14:713–730.

Hoffman FO, Hammonds JS. 1994. Propagation of uncertainty in risk assessments: the need to distinguish between uncertainty due to lack of knowledge and uncertainty due to variability. Risk Anal 14:707–712.

Klir GJ, Yuan B. 1995. Fuzzy sets and fuzzy logic: theory and applications. Upper Saddle River (NJ): Prentice Hall.

Mineau P. 2002. Estimating the probability of bird mortality from pesticide sprays on the basis of the field study record. Environ Toxicol Chem 21:1497–1506.

Norton SB. 1998. A ecological risk assessor's perspective of uncertainty. In: Warren-Hicks WJ, Moore DRJ, editors. Uncertainty analysis in ecological risk assessment. Proceedings of the Pellston Workshop on Uncertainty Analysis in Ecological Risk Assessment, 23 to 28 August 1995. Pensacola (FL): SETAC.

Suter, GW, Barnthouse LW. 1993. Assessment concepts. In: GW Suter II, editor. Ecological risk assessment. Chelsea (MI): Lewis Publishers.

Thayer AM. 1989. Alar controversy mirrors differences in risk perceptions. C&EN Aug 28:7–14.

[USEPA] US Environmental Protection Agency. 1997. Policy for use of probabilistic analysis in risk assessment: guiding principles for Monte Carlo analysis. Washington (DC): ORD, USEPA.

[USEPA] US Environmental Protection Agencyt. 2000. A progress report for advancing ecological assessment methods in OPP: A consultation with the FIFRA Scientific Advisory Panel. Overview document. Available from: http://www.epa.gov/scipoly/sap

Walley, P. 1991. Statistical reasoning with imprecise probabilities. London: Chapman and Hall.

Warren-Hicks WJ, Moore DRJ. 1998. Uncertainty analysis in ecological risk assessment. Proceedings of the Pellston Workshop on Uncertainty Analysis in Ecological Risk Assessment, 23 to 28 August 1995. Pensacola (FL): SETAC.

2 Problem Formulation for Probabilistic Ecological Risk Assessments

A. Hart, S. Ferson, J. Shaw, G. W. Suter II,
P. F. Chapman, P. L. de Fur,
W. Heger, and P. D. Jones

2.1 INTRODUCTION

Problem formulation is an early phase of risk assessment, where the assessment problem is defined and the assessment itself is planned. It provides the foundation for the entire assessment; any deficiencies in problem formulation will compromise all subsequent work on the assessment (USEPA 1998).

Other terms used to describe this phase of the risk assessment process include "problem definition," "problem characterization," "risk profiling" (EC 2000), and "scoping phase."

Extensive general guidance for problem formulation exists already (e.g., USEPA 1998). This chapter reviews the main steps in problem formulation and discusses issues that require special consideration because of the use of uncertainty analysis in probabilistic risk assessment.

2.2 MAIN STEPS IN PROBLEM FORMULATION

The US Environmental Protection Agency (USEPA 1998) describes problem formulation as an iterative process with 4 main components: integration of available information, definition of assessment endpoints, definition of conceptual model, and development of an analysis plan. These 4 components apply also to probabilistic assessments. In addition, it is useful to emphasize the importance of a 5th component: definition of the assessment scenarios. The relationships between all 5 components are depicted in Figure 2.1. Note that the bidirectional arrows represent the interdependency of the different components and imply that they may need to be revised iteratively as the formulation of the problem is refined.

The following sections discuss each of the components of problem formulation in turn, with particular attention to the needs of probabilistic assessments.

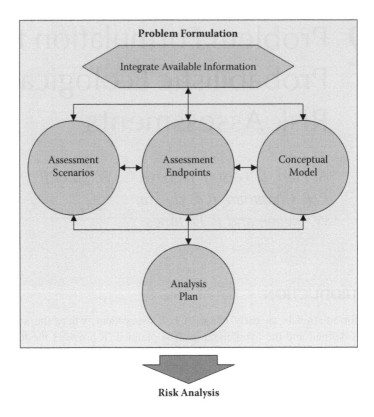

FIGURE 2.1   The main components of problem formulation (adapted from USEPA 1998).

2.3   INTEGRATION OF AVAILABLE INFORMATION FOR PROBABILISTIC ASSESSMENTS

Integration of available information is an iterative process that normally occurs throughout problem formulation (USEPA 1998). In general, for probabilistic assessments there will be a greater emphasis on obtaining information in quantitative rather than qualitative forms. In particular, probabilistic assessments require increased attention to obtaining information on

- Variability
- Uncertainty
- The limits to knowledge
- The quality of studies and data

For example, existing databases and risk assessment publications often omit statistical measures of variability or uncertainty and sample sizes and rarely report the underlying data. These types of information are rarely used in deterministic assessments but are a fundamental requirement for probabilistic assessments.

2.4 DEFINITION OF ASSESSMENT ENDPOINTS FOR PROBABILISTIC ASSESSMENTS

Assessment endpoints are measurable ecosystem characteristics that represent management goals (USEPA 1998). They should define

- The entity to be protected
- An attribute of it that is potentially at risk, important to protect, measurable, and has easily discernible meaning

Often the management goal is not defined in legislation in a specific way, for example, it may refer to prevention of unacceptable adverse effects and not define specifically what is to be protected, nor what measures should be used to represent the magnitude of effects. In this situation, risk assessors and decision makers (or, as sometimes termed, "risk managers") need to agree on assessment endpoints that enable the risk managers to fulfill the requirements of the legislation in an appropriate way.

The assessment endpoint should be not only measurable (at least potentially) but also "modelable." Defining a modelable endpoint is likely to require close discussion between an assessor (who knows what they can model) and a risk manager (who knows what they want to protect). Sometimes the assessment endpoint is only indirectly related to the management goal, for example, if the assessment endpoint is a risk to individuals, but the aim is to protect population sustainability. In such cases, qualitative inference will be required to interpret the assessment result. This inference will need to be done jointly by the risk assessor and risk manager. It is likely to involve substantial uncertainty, which will have to be taken into account qualitatively when producing a narrative description of the assessment outcome. This step should be identified as part of the conceptual model.

If the assessment is to be probabilistic, the risk assessor and risk manager should consider together how this influences the definition of the assessment endpoint. Suter (1998) suggests 5 questions for the risk assessor to ask the risk manager to help define the assessment endpoint:

1) Should any assessment endpoints be expressed as probabilities? Suter (1998) points out that it can be confusing to use the term probability in defining assessment endpoints because it is unclear whether it relates to variability or uncertainty, so it will be helpful to distinguish these in the discussion with the risk manager.
2) If effects are expressed as a threshold value, should the risk be expressed as the magnitude of exceedance, the frequency of exceedance, or the certainty of exceedance?
3) If effects are expressed as some measurement in the field, should they be expressed as the best estimate of the effect, the frequency of exceeding some threshold, or as the certainty that some threshold is exceeded?
4) Would it be useful to know the likelihood that additional data would change the conclusion of the assessment?

5) Should uncertainties other than those concerning the values of parameters simply be listed, listed and scored, or listed and assigned approximate magnitudes based on expert judgment?

Questions 2 and 3 imply a choice between expressing effects in terms of magnitude, frequency, and certainty. In practice, the assessment endpoint may often need to be defined in terms of 2 or 3 of these dimensions. For example, it may be desirable to estimate the proportion of species (frequency) that will experience different levels of mortality (magnitude), and to provide confidence limits (certainty). Indeed, the risk manager's questions may imply an assessment endpoint with more than 3 dimensions, for example, if it is desired to express frequency in terms of space (e.g., number of hectares) and time (proportion of years). The dimensionality of the assessment endpoint will have major implications for all aspects of the analysis and for communication of results, so it is essential to discuss it carefully with the risk manager at the outset to ensure it meets their needs.

If the assessment endpoint is a distribution, or a statistic from a distribution (e.g., 95th percentile), it is essential to be clear how the distribution is interpreted (Suter 1998, p 129). If it is a frequency distribution, to what statistical population does the distribution refer? For example, does the distribution represent a population of individuals, an assemblage of species, a number of locations treated with pesticides, or a series of time periods? The answer to this question has substantial implications for the structure of the assessment model and the types of data required.

2.5   DEFINITION OF ASSESSMENT SCENARIOS

It is essential to define the assessment scenario within which the assessment endpoint will be assessed. The assessment scenario should specify the spatial, temporal, and ecological boundaries within which the endpoint is assessed, since these have substantial implications for the structure of the assessment model and the scope of the input data. The assessment scenario should also describe those aspects of the ecosystem that are relevant to the assessment, that is, those aspects that have an influence on the mechanisms of exposure and effects that will be assessed. This step is important in all ecological risk assessments; it places the assessment activity into the real context of an ecosystem, helps to prevent construction of inappropriate models, and helps with interpretation and communication of results.

The choice of assessment scenario, like the assessment endpoint, is likely to be implied by the management goal and should be made in close consultation with the risk manager, to ensure it meets their needs.

For pesticide risk assessments, it may often be necessary to assess impacts of the same pesticide used in different crops, in different seasons, in different geographic regions, and on different species and ecosystems. This will require the use of multiple scenarios and possibly multiple assessment endpoints.

Multiple scenarios may also be necessary to allow assessment of endpoints at different levels of temporal, spatial, and biological scale (US SAP 1999). This is because both the risks and their acceptability to stakeholders may differ markedly between levels (e.g., the frequency of bird kills versus the risk of sustained population decline).

Another important reason for using multiple scenarios is to represent major sources of variability, or what-if scenarios to examine alternative assumptions about major uncertainties. This can be less unwieldy than including them in the model. Also, the distribution of outputs for each separate scenario will be narrower than when they are combined, which may aid interpretation and credibility. A special case of this occurs when it is desired to model the consequences of extreme or rare events or situations, for example, earthquakes. An example relevant to pesticides might be exposure of endangered species on migration. This use of multiple scenarios in ecological risk assessment has been termed "scenario analysis," and is described in more detail in Ferenc and Foran (2000).

2.6 DEVELOPING CONCEPTUAL MODELS FOR PROBABILISTIC ASSESSMENTS

Conceptual models consist of 2 principal components (USEPA 1998):

- A set of risk hypotheses that describe predicted relationships among stressor, exposure, and assessment endpoint response, along with the rationale for their selection
- A diagram that illustrates the relationships presented in the risk hypotheses

Examples of risk hypotheses given by USEPA (1998) are textual (e.g., "birds die when they consume recently applied granulated carbofuran"). For a quantitative risk assessment it will be preferable to express risk hypotheses using formal mathematical equations.

Various approaches and graphical conventions have been used in drawing conceptual model diagrams. Consideration could be given to recommending a standardized approach for use in probabilistic assessments.

Suter (1999) makes a number of recommendations that may be helpful:

- That conceptual model diagrams be constructed as a cascade of alternating processes and states
- That exposure–response relationships be shown as distinct components of model diagrams
- That more complex problems be represented by a hierarchy of conceptual models, with each lower level containing states and processes that are aggregated at the next higher level
- That conceptual models be developed in a modular way, standard modules being developed to represent states and processes that occur repeatedly in many assessments

In addition, it may be useful to extend the conceptual model to show the relationships between the modeled states and processes and the types of information that will be used to quantify them. This is useful because, generally, model parameters are not measured directly but are estimated from other information. Including this in the conceptual model makes the extrapolation explicit, and recognizes the attendant uncertainty.

It may also be useful to include in the conceptual model other lines of evidence that are relevant to the assessment endpoint. This may help to highlight the contribution that other lines of evidence can make and promote more effective gathering and use of such information.

2.6.1 DEFINING AN APPROPRIATE STRUCTURE FOR THE CONCEPTUAL MODEL

Some important criticisms encountered by pioneering efforts to apply probabilistic methods to pesticides could have been avoided by appropriate structuring of the conceptual model. These criticisms have included

- Exposure should usually be evaluated and effects predicted initially for individuals, not at higher levels (US SAP 1999).
- Inappropriate averaging and/or aggregation of exposure or effects to higher levels can create misleading results (US SAP 1999).
- For each probabilistic component of an analysis, it is essential to be clear what is distributed, and with respect to what variable it is distributed, and to ensure that the data used for the distribution are consistent with that interpretation (Suter 1998, p 129–130).
- When distributions are combined, for example, in joint probability curves, it is important to ensure that the resulting distribution is meaningful, again in terms of what is distributed and with respect to what variable (Suter 1998, p 129).

To help address these issues, we define a new component for use in conceptual models: the "units of analysis." These are the lowest levels of biological, spatial, and temporal scale used in the quantitative part of the risk assessment (e.g., individual iterations in a simulation model). They also define the biological, spatial, and temporal units of the measures that will be needed as inputs to the assessment model.

The relationships between this concept and other concepts used in the USEPA's Risk Assessment Framework (USEPA 1998) are illustrated in Figure 2.2. The arrows indicate how the elements are related. The management goal defines the assessment endpoint. The nature and dimensionality of the assessment endpoint in turn defines the units of analysis, i.e., the lowest levels of biological, spatial, and temporal scale that need to be identified in the assessment model. The units of analysis represent real units of scale in real-world processes of exposure and effects, e.g., individual animals, their ranges of movement, their daily food intakes, and their life spans. Measurements of relevant quantities in the real world are used to estimate the measures that are processed by the risk assessment model to produce a risk estimate. If the model aggregates the units of analysis correctly, the risk estimate will be an estimate of the assessment endpoint. The risk estimate will often be used to produce a textual risk description; these two together are then used to inform the management

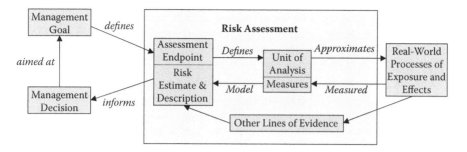

FIGURE 2.2 How the unit of analysis relates to other components of the assessment process.

decision, which is aimed at achieving the management goal. The diagram also includes other lines of evidence. These may be other types of information about exposure and effects in the real world. They may be incorporated quantitatively in the risk estimate (e.g., by Bayesian updating) or subjectively in the risk description (e.g., by weight of evidence).

The criticisms listed at the start of this section should be avoided by

- Defining the units of analysis at appropriate levels
- Ensuring the units of analysis represent real-world processes appropriately
- Defining an appropriate relationship between the units of analysis and the assessment endpoint

The following sections discuss how to define the unit of analysis, and how to define the relationship between the unit of analysis and the assessment endpoint.

2.6.2 DEFINING THE UNITS OF ANALYSIS

The units of analysis should be determined by the needs of the assessment, not by the data that happen to be available. Careful consideration is required to identify which biological, spatial, and temporal units are appropriate for each assessment. This will depend on the nature and degree of spatial and temporal variation in the many factors that affect exposure and effects, including the following:

- The source of the stressor
- The behavior or fate of the stressor in the environment
- The behavior of the receptor organisms in the environment
- The relationship between the temporal pattern of exposure and the level of effects for receptor organisms

It has been argued that exposure of birds should usually be evaluated and effects predicted initially for individuals, and then used to evaluate consequences at larger scales (US SAP 1999), because it is individuals that experience mortality or fail to reproduce. Also, spatial and temporal variation in pesticide residues combined with variation in the foraging behavior of individual birds means that a single pesticide application event results in widely varying exposure for different individuals. In this situation, it may be appropriate to define the biological unit of analysis at an individual level, as suggested by the USEPA Science Advisory Panel (US SAP).

In some assessments it may be reasonable to assume that all individuals are affected in the same way. For example, it is usually assumed that all fish in a water body are exposed to the same concentration of pesticide. In this case, it is unnecessary to model the exposure of each individual; modeling the group as a whole is simpler and will give the same result.

In principle avian exposure could be modeled at very fine levels of spatial and temporal scale: e.g., estimating residues of pesticides on individual seeds and insects and then modeling individual choices of a bird feeding on them — an analysis in units of centimeters and seconds. This level of analysis is very cumbersome, and usually unnecessary. If the results would be the same, the analysis may be done at larger scales (e.g., in units of fields and days).

However, a finer level of detail may be required in some situations. For example, predatory birds feeding on rodents in an area partly treated with rodenticides may encounter a bimodal distribution of residues in their prey, such that most prey contain no residues but others contain a lethal dose for the predator. If the unit of analysis were defined as a whole day's foraging, with residues being averaged over all available prey items, the model might indicate that all the predators experience a sublethal exposure. In reality, most predators would experience zero exposure, but those that ate a contaminated prey item would die: a significantly different result. Therefore, the unit of analysis in this case should be individual foraging events for individual predators.

These examples illustrate the care that is required in defining units of analysis, and suggest 2 general principles. First, defining the unit of analysis at a higher level is inappropriate if including variability or uncertainty at a lower level would give a different result. Second, it is desirable to avoid defining the unit of analysis at a lower level than necessary (an example of "Occam's razor"). Vose (2000, p 203) offers guidance that incorporates both these principles: "a model should be disaggregated as much as is necessary ... to express any significant logic between input variables; and to model each uncertain variable as accurately as is necessary for the efficient but accurate modeling of the problem." Vose also suggests that models should be evenly disaggregated, i.e., the variables should be broken down so that each component has about the same effect on the uncertainty of the output. This seems reasonable, provided that the 1st principle is not violated.

In defining units of analysis it is also important to take account of Vose's (2000, p 201) cardinal rule of risk analysis modeling: "every iteration of a risk analysis model must be a scenario that could physically occur."

Finally, it is essential that the units of the analysis are suitable for generating the assessment endpoint.

Tabulating the temporal, spatial, and biological scales of each component of the assessment may help to identify appropriate units of analysis, show how they relate to real-world processes, and check their compatibility with the assessment endpoint and hence the management goal (e.g., Table 2.1).

TABLE 2.1

A tabular approach to identifying appropriate biological, spatial, and temporal scales for different components of the assessment process, illustrated for a hypothetical assessment of risks to birds from a corn insecticide (see also Figure 2.2)

	Management goal	Assessment endpoint	Unit of analysis	Real-world processes
Biological entity	Birds of all the species present in corn-growing areas	Local populations of single "focal" species, chosen to represent a range of similar species around cornfields	Individual of the "focal" species	Individual birds in a number of species with differing behavior and ecology
Biological attribute	"No unreasonable adverse effects" (FIFRA); no regularly repeated bird kills (USEPA 1998)	Likelihood, frequency, and magnitude of bird kills caused by pesticide exposure	Mortality (yes or no)	Intoxication and mortality or recovery
Spatial scale	Corn-growing regions	Cornfields and adjacent habitats in a representative corn-growing region	Individual's feeding range; divided into corn, drift zone, and other habitat	Centimeters (insects from different locations within cornfield canopy carry different residues)
Temporal scale	Years	1 year	1 day	Seconds (interval between successive foraging choices)

2.6.3   DEFINING THE RELATION BETWEEN UNITS OF ANALYSIS AND THE ASSESSMENT ENDPOINT

The management goal will often require that the assessment endpoint is defined at higher levels of biological, spatial, and temporal scales than the units of analysis. In some cases, it may be appropriate to generate the assessment endpoint by simple averaging or aggregation of the units of analysis, but in other cases this may require modeling of complex population processes.

Inappropriate averaging or aggregation creates misleading results. For example, averaging exposure over space, time, and species before predicting effects will dilute higher levels of exposure and can cause gross underestimation of effects (US SAP 1999).

Careful construction of the conceptual model diagram, and the use of a tabular approach such as Table 2.1, should help to avoid these problems. The diagram should show clearly the point at which individual exposure is used to predict individual effects and the process by which individual effects are aggregated to generate the risk estimate. In addition, it should be remembered that the risk estimate may be combined quantitatively or qualitatively with other lines of evidence to address the assessment endpoint.

2.6.4   IDENTIFYING UNCERTAINTIES IN THE CONCEPTUAL MODEL

Two common failings of probabilistic assessments (Warren-Hicks and Moore 1998; US SAP 1999) are

- Failure to identify and address key uncertainties
- Failure to identify and include dependencies

It is useful to distinguish between variability, parameter uncertainty, and model uncertainty, since they require different treatment in risk analysis (Suter and Barnthouse 1993). Variability refers to actual variation in real-world states and processes. Parameter uncertainty refers to imprecise knowledge of parameters used to describe variability or processes in a risk model; this can arise from many sources including measurement error, sampling error, and the use of surrogate measurements or expert judgment. Model uncertainty refers to uncertainty about the structure of the risk model, including what parameters should be included and how they should be combined in the model equations.

Model uncertainty is often overlooked. It results when there is disagreement within the scientific community about the underlying processes, when the underlying mechanisms are poorly characterized, when extrapolation beyond existing data or theory is necessary, or when the conceptual model is poorly formulated. Examples include

- Lack of knowledge about how the ecosystem functions
- Omission of relevant stressors, routes of exposure, types of effect
- Overlooking secondary effects
- Incorrect boundaries for the assessment (spatial, temporal, or biological)

- Inappropriate assumptions about representation, extrapolation, or functional forms
- Inappropriate selection, averaging, or aggregation of variables
- Failure to identify and interrelate temporal and spatial parameters
- Overuse or misuse of independence assumptions

Useful strategies for identifying and characterizing uncertainties:

- Systematically examine each risk hypothesis for each type of analytical uncertainty, include them in the description of the risk hypothesis, and state in which direction they are expected to affect the assessment endpoint.
- Prioritize or rank the uncertainties affecting each risk hypothesis.
- Show at least the major uncertainties in the diagram of the conceptual model, linked to the model components they affect (e.g., Figure 2.3).
- Rank the model components in terms of their uncertainty.
- Try drawing alternative model diagrams to identify structural uncertainties.
- Produce a summary description of nature of uncertainties at the close of problem formulation.

2.6.5 IDENTIFYING DEPENDENCIES IN THE CONCEPTUAL MODEL

Dependencies among the input variables of a risk model can have pronounced effects on the output distribution, especially in the tails (Warren-Hicks and Moore 1998; US SAP 1999). Rainfall is fully independent of the intrinsic chemical properties of the pesticide, so that neither one depends on the other. But field conditions will most certainly affect the fate and transport of a pesticide once it is applied to the field. For example, the evaporation of the chemical from the field or plant surface depends on ambient temperature. Types of dependency include the familiar cases of independence and linear correlation, but also more complex relationships (Figure 2.4).

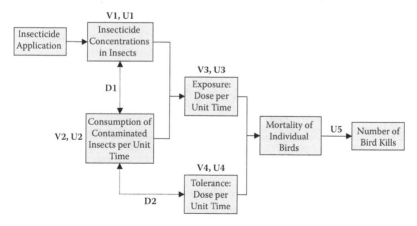

FIGURE 2.3 A simple approach to identifying variables, uncertainties, and dependencies in a conceptual model diagram. For key, see Table 2.2.

TABLE 2.2
Key to variables (V), uncertainties (U), and dependencies (D) in Figure 2.3
(Note that the lists are illustrative, not exhaustive)

V1   Variation over space and time in the concentrations of pesticide in insects

U1   Extrapolation from field studies with other pesticides, and/or sampling uncertainty due to limited numbers of field sites

V2   Intraspecies variation in food requirement and dietary composition; variation between individuals and over time in the proportion of food that is contaminated

U2   Uncertainty in estimating food intake from body weight and energy content of food; assumptions in estimating proportion of food that is contaminated

V3   Variation in dietary exposure between individuals and over time (due to V1 and V2)

U3   Uncertainty about the contribution of nondietary routes of exposure (assumed in the model to be zero)

V4   Intraspecies variation in toxicological sensitivity

U4   Uncertainty in extrapolating toxicity from laboratory species to focal species; uncertainty in estimating intraspecies variation

U5   Uncertainty about relationship between individual mortality rate and number of "kills" (local episodes of mortality)

D1   Intake may be negatively correlated with residues due to repellency or avoidance; positive relation between insect abundance and decision to use insecticide; negative relation between use of insecticide and subsequent insect abundance; positive relation between insect abundance and bird foraging behavior

D2   Food consumption and tolerance are both functions of body weight

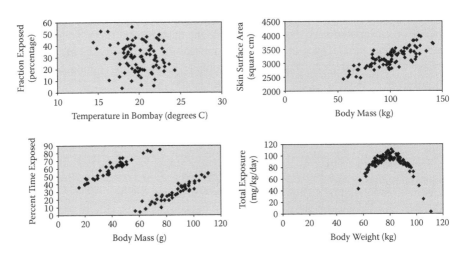

FIGURE 2.4 Examples of four types of dependency. In the bottom left-hand graph the 2 groups of points might represent males and females, for example, including random, linear, parallel, and curvilinear relationships.

Perhaps most easy to overlook are spatial and temporal dependencies. For example, the hydrologic component of the pesticide root zone model–exposure analysis modeling system (PRZM–EXAMS) treats multiple field plots over whole watersheds as independent, uncoupled, simple, 1-dimensional flow systems. In reality, the field plots are coupled systems that exhibit complex 3-dimensional water flow and pesticide transport (US SAP 1999). These higher order processes introduce spatial dependencies that may need to be considered in the assessment. Temporal autocorrelations are also likely when assessing exposure.

To reduce the risk of overlooking dependencies, it may be useful to

- Systematically examine all model components for possible interdependencies
- Describe the form of each dependency and identify which risk hypotheses it affects
- Show major interdependencies in the model diagram (e.g., Figure 2.3)

When using empirical data to check for dependencies, it is important to remember that they are hard to measure, especially when sample sizes are low. In addition, zero correlation does not necessarily imply independence, and pairwise independence does not imply mutual independence, since more complex dependencies may be present.

2.7 ANALYSIS PLANS FOR PROBABILISTIC ASSESSMENT

As risk assessments become more complex, the importance of a good plan increases (USEPA 1998). The plan should identify

- Which risk hypotheses will be assessed
- Which new and existing data will be used
- What methods of analysis will be used
- How uncertainties will be dealt with
- Whether and how the analysis will be phased or tiered

When planning probabilistic assessments, the following issues require special attention.

2.7.1 SELECTION AND PARAMETERIZATION OF DISTRIBUTIONS

A critical extra phase to be included when planning probabilistic assessments is the selection and parameterization of distributions, to represent the sources of variability and uncertainty that have been identified in the conceptual model. The issues and approaches involved are discussed elsewhere in this book.

The resource consumed by this activity may be reduced if standard distributions can be adopted for parameters that are required for many different assessments. However, caution should be exercised to avoid applying default distributions outside the range of problems for which they are appropriate.

2.7.2   Propagating Variability and Uncertainty

A 2nd critical addition when planning a probabilistic assessment is the choice of methods for propagating variability and uncertainty. The workshop reviewed a range of contrasting methods of analyzing uncertainty in risk assessments:

- Interval analysis
- Probability bounds analysis
- First-order error analysis
- First-order (nonhierarchical) Monte Carlo
- Second-order (hierarchical or 2-dimensional) Monte Carlo
- Bayesian methods

From the standpoint of practical regulatory assessment, it would be desirable to reach a consensus on the selection of methods for routine use for pesticide risk assessments while recognizing that there may be scientific reasons for preferring alternative methods in particular cases. Such a consensus does not yet exist. Further case studies are required, covering a range of contrasting pesticides and scenarios, to evaluate the available methods more fully. While a consensus is lacking, it is important that reports on probabilistic assessments clearly explain how their methods work and why they were selected.

2.7.3   Separation of Variability and Uncertainty

An important question when planning a probabilistic assessment is whether to separate variability and uncertainty in the analysis and results. This is one of the "key issues" that were given special consideration at the Pellston workshop that developed this book. While there was not a consensus, the majority view was that there are potential advantages to separating variability and uncertainty, but further case studies are needed to evaluate the benefits and practicality of this for routine pesticide assessment.

2.7.4   Dealing with Dependencies

Another important question when planning a probabilistic assessment is how to deal with dependencies. This also is one of the "key issues" that were identified for the Pellston workshop. Further work is needed to evaluate these options. Some additional points are made here.

Even if a correlation is below the conventional level of significance, consideration should be given to whether it might alter the risk estimate, and it may be prudent to include it. When measured or estimated correlations are used to specify dependencies in Monte Carlo models, it is important to check that the matrix of correlations satisfies mathematical constraints (Table 2.3).

If there are significant spatial or temporal dependencies, it may be necessary to use a spatially or temporally explicit model in order to avoid misleading results.

TABLE 2.3
Correlation coefficients used to define dependencies in risk models must be positive–definite: e.g., for the variables *A, B,* and *C,* if the correlations *r* and *s* are big and positive, *t* must be positive too

	A	B	C
A	1	r	s
B	r	1	t
C	s	t	1

2.7.5 DEALING WITH MODEL UNCERTAINTY

The workshop recognized the importance of dealing with model uncertainty but did not evaluate the alternative approaches in detail. Further work is required to identify instances of model uncertainty for pesticide risk assessment and to develop guidance on how to deal with it. Some possible approaches are briefly discussed below.

2.7.5.1 Model Weighting

Model uncertainty can be represented by formulating 2 or more different models to represent alternative hypotheses or viewpoints and then combining the model outputs by assigning weights representing their relative probability or credibility, using either Bayesian and non-Bayesian approaches.

Model weighting is considered inappropriate and misleading by many. Morgan and Henrion (1990) point out that while we may be able to say 1 model produces more accurate predictions than another, we cannot say that 1 model is more probable than another because, ultimately, every model is definitely false. Combining models probabilistically using Monte Carlo simulation treats uncertainty as though it were variability. It may also have the effect of averaging together 2 different theories to generate an outcome that is actually compatible with neither. For example, suppose we don't know whether a widely distributed substance is carcinogenic. Model weighting treats uncertainty about the carcinogenicity as though it were variability in the carcinogenicity. But the situation of having either a lot of people getting cancer or none getting cancer is very different from having a moderate number of people get cancer, or a lot of people getting, say, nonmalignant tumors.

2.7.5.2 Scenario Analysis

A conceptually simple approach that avoids the difficulties of model weighting is scenario analysis or the "1-at-a-time" method, where the alternative models are analyzed separately and the results are compared. In the example of the previous section, this might produce a conclusion of the type "If model A is true then 0 people will get cancer; if model B is true then 200 people will get cancer." However, this approach becomes laborious to implement and complex to interpret when many separate aspects of the model structure are uncertain, requiring large numbers of alternative models and multiple comparisons of outputs.

2.7.5.3   Model Enveloping

Multiple comparisons of alternative models can be avoided by use of model enveloping. This can be done relatively simply using bounding methods such as probability bounds analysis. Continuing the example of the previous sections, model enveloping might produce a conclusion such as "Between zero and 200 people will get cancer." However, this can still give the impression that all intermediate points are possible when in fact some of them are compatible with none of the competing theories.

2.7.5.4   Meta-Models

Another approach is to develop a global model that contains plausible models as special cases, defined by alternative values of particular parameters. This converts model uncertainty into uncertainty about the model parameters. Again this can be done using either Bayesian or non-Bayesian approaches. This approach is favored by Morgan and Henrion (1990), who describe how it can be applied to uncertainty about dose–response functions (threshold versus nonthreshold, linear versus exponential).

2.7.6   Sensitivity Analysis

Some sources of uncertainty and variability may have so little influence on risk that they can be held constant and not treated probabilistically in the assessment. The analysis plan should state the rationale for deciding which variables and hypotheses this applies to (USEPA 1998).

Sensitivity analysis provides a good tool for this purpose (USEPA 1997). It quantifies the change in model outputs as a function of changes in each model input and enables the influence of different inputs to be compared.

Different methods of sensitivity analysis will produce different results, so they should be chosen carefully (Warren-Hicks and Moore 1998). A comprehensive account of alternative approaches is provided by Saltelli et al. (2000).

As well as guiding problem formulation, sensitivity analysis can be valuable in optimizing the use of resources. By revealing which uncertainties have the most influence on the results of the assessment, sensitivity analysis can also help target additional research or monitoring; and by revealing which of the controllable sources of variability have the most influence, sensitivity analysis can help identify and evaluate practical options for managing risk.

2.7.7   Incorporating Other Lines of Evidence

Probabilistic models will normally not be the sole basis for decision making but will be considered together with other lines of evidence (see Figure 2.2). The way in which this will be done should be considered at the outset as part of problem formulation and specified in the analysis plan.

Before deciding to treat a line of evidence separately, consideration should be given to whether it can in fact be directly incorporated into the quantitative assessment. For example, it may be possible to use Bayesian updating to incorporate information from field studies or monitoring if they provide direct measurements of the assessment endpoint, or of the intermediate model.

Usually, some lines of evidence will not be suitable for direct incorporation into quantitative analysis. Semiquantitative or qualitative methods will then be needed to weigh the different lines of evidence, including the quantitative assessment, and integrate them for decision making. Methods for assessing weight of evidence were outside the scope of the workshop that developed this book but are discussed by Suter et al. (2000) and were recently the focus of another workshop (Chapman et al. 2002). Whatever method is used for weighing different lines of evidence, it will be important to characterize uncertainties in each line of evidence and show their effect on the overall assessment outcome.

2.7.8   How to Present the Results

The analysis plan should specify not only how the analysis will be conducted, but also how the results will be presented. Indeed, the way results will be communicated will usually influence the choice of both model structure and analysis method and is ultimately driven by the information needs of risk managers and other stakeholders and their management goals (see Figure 2.2). Careful advance planning for the communication of results is especially important for probabilistic assessments because they are more complex than deterministic assessments and less familiar to most audiences. It may be beneficial to present probabilistic and deterministic assessments together, to facilitate familiarization with the newer approaches.

2.7.9   Tiering the Risk Assessment Process

Tiers are widely used to help improve the efficiency of the risk assessment process. They generally start with a simplified assessment to screen out scenarios with obviously high or low risks.

An approach that is sometimes used in deterministic assessments is to set some variables to "realistic worst case" values in early tiers and gradually introduce more "typical" (but still deterministic) values in later tiers. Some current proposals for probabilistic assessment of pesticides have adopted a modified version of this approach, suggesting that worst case assumptions should gradually be replaced with distributions as the assessment is refined (ECOFRAM 1999; USEPA 2000). This makes the assessment output conservative (i.e., tending to overestimate risk) because it refers to conditions that are partly worst case. However, the degree of conservativism will be unclear. Nevertheless, this approach may still provide sufficient information for a decision to be made, if the results show that the risk is acceptable despite the bias toward worst case. The main advantage of this approach is that it requires less time and resources than would be needed to quantify all sources of variability and uncertainty. However, others have argued that no conservative assumptions should be included in a probabilistic assessment and that the proper place for conservatism is in the risk manager's decision process (Moore et al. 1999). Which of these approaches is appropriate is an important issue that may vary between jurisdictions and assessments depending on the objectives of the risk managers.

Optimizing the use of probabilistic methods within the regulatory assessment process, and especially within tiered assessments, was recognized as one of the "key issues" that were given special consideration at the Pellston workshop that developed this book.

2.7.10   DECIDING WHEN TO STOP THE ASSESSMENT

To avoid wasting resources by overrefining assessments, avoid "paralysis by analysis," and reassure stakeholders that the assessment process is finite, criteria are needed for deciding when to stop.

A risk assessment can be considered complete when risk managers have sufficient information and confidence in the results of the risk assessment to make a decision (either positive or negative) that they can defend (USEPA 1998). Identifying when this point is reached will require repeated consultation between risk assessors and risk managers as the assessment progresses, unless they can define stopping rules in advance. Such stopping rules would need to specify what frequency and magnitude of impact is acceptable, and also define acceptable limits on decision errors (equivalent to Step 6 of the USEPA's Data Quality Objectives [DQO] process; Suter 1998). The Ecological Committee on FIFRA Risk Assessment Methods (ECOFRAM 1999) proposed defining a "threshold of acceptable risk" as a stopping rule.

Defining stopping rules in advance is undoubtedly difficult: in 1998 Suter wrote that Step 6 of the DQO process had never been completed for an ecological risk assessment. Unless this obstacle can be overcome, frequent consultation between risk assessors and managers will be needed during each assessment to avoid overrefining it (ECOFRAM 1999).

2.7.11   NEED FOR DIALOGUE

It is essential to have a clear vision of roles of the different parties to risk assessment, including risk assessors, risk managers, and other stakeholders, and to ensure they interact efficiently throughout the process (National Research Council 1983, 1996; US Presidential/Congressional Commission on Risk Assessment and Risk Management 1997).

This need for dialogue applies to all phases of the risk analysis process, including problem formulation. The parties to a risk assessment need to communicate and cooperate during problem formulation and agree on each component (see Figure 2.1) before proceeding to the analysis phase (USEPA 1998; US SAP 1999). The need is greater for probabilistic assessments because they are less well established and more complex than deterministic assessments. Without dialogue, there is a high risk of mistakes, disagreements, and wasted effort.

2.7.12   GENERIC PROBLEM FORMULATIONS

It will be apparent from this chapter that problem formulation for a probabilistic assessment can be a substantial undertaking, and perhaps the most difficult and critical part of the whole process. Therefore, there would be great benefits if it were possible to create "generic" problem formulations, or generic components for problem formulation, that were genuinely appropriate for a number of different assessments without having to repeat the whole process.

The prospects for creating such generic assessments are good, at least for pesticides. There is a high level of consistency in the assessment endpoints and scenarios that are relevant for different pesticides with similar use patterns (especially if they have similar chemistry and mode of action), and therefore a high consistency in the conceptual models and analysis plans that are appropriate. This is reflected in the high level of standardization that is typical of current deterministic pesticide assessment in both Europe and North America and contrasts with the more case-by-case approach that is necessary for contaminated land assessments (e.g., Superfund). The tiered assessment procedures laid out in guidance documents and regulations for pesticides, and the use of standardized computer models such as PRZM–EXAMS, imply generic problem formulations even if they are not described as such. The tiered approach for estimating pesticide concentrations in surface waters, recently developed in Europe, explicitly defines 9 different assessment scenarios (FOCUS 2001).

Creating generic problem formulations for probabilistic assessments is, because of their greater complexity, both more challenging and more worthwhile. The USEPA has already begun this process for both aquatic and terrestrial pesticide risks (USEPA 2000). More generally, consideration has been given to the development of modular conceptual models for complex ecological assessments (Suter 1999) and of generic assessment endpoints (USEPA 2003).

It is important to note that if generic problem formulations are to be used, it is especially important that they are developed very carefully in the first place, that their domains of applicability are carefully defined, and that users should double-check on every occasion that they are fully appropriate to the case in hand or adjust them as necessary.

2.7.13 DOCUMENTING PROBLEM FORMULATION

Finally, it is essential to document all components of problem formulation fully and clearly, so that the basis of every assessment is explicit and open to review, and so that assessments can if necessary be duplicated by different assessors. For the sake of transparency and public trust, it may also be desirable to document the process that developed the problem formulation, including the nature of interactions between risk assessor and risk manager and the involvement of stakeholders. Recording all this information clearly without generating unmanageable amounts of documentation may be a challenge, but was not discussed at the workshop that produced this book. Figure 2.3 and Tables 2.1 and 2.2 might be useful starting points for developing some types of summary presentation, but it will also be important to express every model using formal mathematical equations. Further work is required to develop effective approaches for both developing and communicating problem formulation for probabilistic assessments.

2.8   REFERENCES

Chapman PM, McDonald BG, Lawrence GS. 2002. Weight-of-evidence issues and frameworks for sediment quality (and other) assessments. Human Ecol Risk Assess 8:1489–1515.

[EC] European Commission. 2000. First report on the harmonisation of risk assessment procedures. Part 1. European Commission, Health and Consumer Protection Directorate-General, Brussels: EC.

[ECOFRAM] Ecological Committee on FIFRA Risk Assessment Methods. 1999. ECOFRAM Terrestrial Draft Report. Ecological Committee on FIFRA Risk Assessment Methods.

Ferenc S, Foran J, editors, 2000. Multiple stressors in ecological risk and impact assessment: approaches to risk estimation. Pensacola (FL): SETAC and Chemistry Press.

[FOCUS] Forum for Coordination of Pesticide Fate Models and Their Use. 2001. Report prepared by the FOCUS Working Group on Surface Water scenarios: "FOCUS Surface Water scenarios in the EU evaluation process under Dir. 91/414/EEC," SANCO/4802/2001-rev. 2 final May 2003, p 238.

Moore DRJ, Sample BE, Suter GW, Parkhurst BR, Teed RS. 1999. Risk-based decision making: the East Fork Poplar Creek case study. Environ Toxicol Chem 18:2954–2958.

Morgan MG, Henrion M. 1990. Uncertainty: a guide to dealing with uncertainty in quantitative risk and policy analysis. Cambridge (UK): Cambridge University Press.

National Research Council. 1983. Risk assessment in the federal government: managing the process. Washington (DC): National Academy Press.

National Research Council. 1996. Understanding risk: informing decisions in a democratic society. Washington (DC): National Academy Press.

Saltelli A, Chan K, Scott EM. 2000. Sensitivity analysis. Chichester (UK): John Wiley & Sons Ltd.

Suter GW II. 1998. An overview perspective of uncertainty. In: Warren-Hicks WJ, Moore DJ, editors. Uncertainty analysis in ecological risk assessment. Pensacola (FL): SETAC, p 121–130.

Suter GW II. 1999. Developing conceptual models for complex ecological risk assessments. Human Ecol Risk Assess 5:375–396.

Suter GW II, Barnthouse LW. 1993. Assessment concepts. In: Suter GW II, editor. Ecological risk assessment. Chelsea (MI): Lewis Publishers.

Suter GW II, Efroymson RA, Sample BE, Jones DS. 2000. Ecological risk assessment for contaminated sites. Boca Raton (FL): Lewis Publishers.

[USEPA] US Environmental Protection Agency. 1997. Policy for use of probabilistic analysis in risk assessment: guiding principles for Monte Carlo analysis. Washington (DC): ORD, USEPA.

[USEPA] US Environmental Protection Agency. 1998. Guidelines for ecological risk assessment. Washington (DC): USEPA.

[USEPA] US Environmental Protection Agency. 2000. Technical Progress Report: Implementation Plan for Probabilistic Ecological Assessments. Washington (DC): Environmental Fate and Effects Division, USEPA.

[USEPA] US Environmental Protection Agency. 2003. Generic ecological assessment endpoints (GEAEs) for ecological risk assessment. Risk Assessment Forum. Washington (DC): USEPA.

US Presidential/Congressional Commission on Risk Assessment and Risk Management. 1997. Final Report, Volume 1. Washington (DC): GPO.

[US SAP] USEPA Science Advisory Panel. 1999. Report of the FIFRA Scientific Advisory Panel Meeting, September 23, 1999. Session V. SAP Report No. 99-05E. Washington (DC): US SAP.

Vose D. 2000. Risk analysis: a quantitative guide. 2nd ed. Chichester (UK): John Wiley & Sons.

Warren-Hicks WJ, Moore DJ, editors, 1998. Uncertainty analysis in ecological risk assessment. Pensacola (FL): SETAC.

3 Issues Underlying the Selection of Distributions

D. Farrar, T. Barry, P. Hendley, M. Crane,
P. Mineau, M. H. Russell, and E. W. Odenkirchen

3.1 INTRODUCTION

To develop a probabilistic model, one has to assign probability distributions to model inputs such as degradation rates, partition coefficients, dose–response parameters (or dose–time–response parameters), exposure values, and so on, for a model relating impacts to exposure. This chapter is concerned with several kinds of technical decisions involved in the selection of distributions.

The simplest situation is represented by most 1-dimensional (1D) models in which the distributions are taken to represent variability, and where there are adequate data to characterize the distributions. More complicated situations may involve 1D modeling with data that are inadequate or problematic (e.g., because of availability of only summary statistics), or the inclusion of uncertainties in 2-dimensional (2D) models.

For distributions that represent variability, initial decisions may relate to the selection of data on which to base distributions. The problem formulation must identify meaningful populations. Ideally, the data are a random sample from the populations of interest; in practice, one may be happy to establish that the data are representative. In addition, data should represent a spatiotemporal scale appropriate for the model.

Having selected an appropriate data set, we must select a type of distribution and fit the distribution to the data, or else use an empirical or other nonparametric distribution. There appears to be some mechanistic basis for the log-normal distribution, for environmental concentrations (Ott 1990, 1995). However, in a given situation there may not be very strong theoretical support for a specific type of distribution, log-normal or otherwise. Alternative distributions may need to be considered based on the quality of fit of the distribution to data. Therefore, it is desirable to have quantitative indices that can be used to compare or rank distributions based on agreement with data. The fit of the log-normal distribution (or whatever distributions we may choose) should be evaluated in particular situations, using graphical as well as statistical procedures.

An alternative to choice of a parametric distribution is to rely on a "nonparametric" distribution. The simplest such distribution is the "empirical" distribution, which assigns equal probability to each datum in a specified dataset. Considerable discussion among environmental risk assessors has focused on use of empirical versus parametric distributions.

Regarding the distribution-fitting step, a good point of departure is the 2-parameter log-normal distribution. The distribution has a probability density function (pdf) of the following form:

$$f(x) = \frac{1}{x} \frac{1}{\sigma\sqrt{2\pi}} \exp\left(-\frac{1}{2\sigma^2}\left(\log(x) - \mu\right)^2\right)$$

(3.1)

Here x represents a specific value of the "distributed" (or random) variable. The quantities μ and σ are the parameters of the distribution. In this case the 2 parameters are interpreted as the mean and standard deviation of the logarithms of the variable. However, distributions in general do not need to be represented in terms of parameters that include a mean and variance in any scale.

If we are to use a log-normal distribution (or any other parametric distribution), values have to be assigned to the parameters, based on data or some rational argument. For the log-normal distribution, given the characterization of μ and σ as log-scale mean and standard deviation, an obvious approach is to transform values in some suitable dataset to logarithms and use the sample mean (of the logarithms) to estimate μ, and sample standard deviation to estimate σ. However, as for distributions of many types, there is more than 1 reasonable approach for estimating log-normal parameters. Below, a brief account is provided of estimation procedures and criteria for evaluation of estimation procedures.

A probabilistic model will typically require distributions for multiple inputs. Therefore, it is necessary to consider the "joint" distribution of multiple variables as well as the individual distributions, i.e., we must address possible dependencies among variables. At least, we want to avoid combinations of model inputs that are unreasonable on scientific grounds, such as the basal metabolic rate of a hummingbird combined with the body weight of a duck.

In practice, various complications may be encountered for which the simplistic description above will not be adequate. First, still within the realm of 1D variability modeling, the measurements may be in some sense partially missing, e.g., censored or available only as summary statistics. In addition, methods may be applicable for specifying distributions based on professional judgment, particularly where the probabilities of interest do not represent relative frequencies, or the probabilities of interest do represent relative frequencies, but there are inadequate data to justify particular distributions.

This chapter is structured as follows. Section 3.2 provides a refresher on some principles of distribution theory and estimation theory. The approach is didactic, and practical issues are put off until Section 3.3. Concepts such as skewness and kurtosis are reviewed, useful for characterizing and comparing different distribution types. Some special distributions are mentioned, which are possibly useful in environmental risk assessment.

Also in Section 3.2, several estimation procedures are defined, such as method of moments (MOM), maximum likelihood (ML), and least squares (LS). Criteria are reviewed that can be used to evaluate and compare alternative estimators.

In Section 3.3, the background material developed in Section 3.2 is used in a discussion of practical issues involved in the selection of distributions, particularly for models of pesticide ecological risk. The topics discussed include data representativeness, preliminary data exploration, selection of distribution type, estimation of distribution parameters (distribution fitting), and evaluation of distribution fit.

Finally, Section 3.4 discusses a range of procedures that may be applicable in situations where the information available is less than one would like. The data available may be too few, subject to various kinds of censoring or absence, or available only in summary form.

3.2 TECHNICAL BACKGROUND

3.2.1 UNDERSTANDING DISTRIBUTIONS

3.2.1.1 Characterizing Distribution Shape in Terms of Skewness and Kurtosis

Indices of distribution central tendency and spread are not reviewed here (see Vose 2000, Section 3.2.1). The concept of skewness of a distribution relates to deviations from symmetry of the pdf. The normal distribution has a skewness of zero (the distribution is symmetric, with the familiar bell-shaped pdf). For a distribution with positive skewness the right tail of the distribution is more extended than the left tail; a distribution with the left tail more extended has negative skewness. In many cases, it seems that skewness is associated with a constraint on the permissible values of a variable (Vose 2000, Section 6.7). The idea is that the distribution tail can be more extended in the direction opposite to a bound than in the direction of the bound.

Some literature has defined kurtosis in terms of pdfs that are relatively flat versus relatively peaked at the mode. A tendency in more recent literature is to emphasize the idea of "tail weight." "Leptokurtic" distributions have relatively heavier pdf tails, while platykurtic distributions have relatively lighter tails (Balanda and MacGillivray 1998).

Leptokurtic distributions are more "outlier-prone." When fitting distributions to data, it may sometimes be difficult to decide whether one should assume a leptokurtic distribution (say, a Student t distribution with relatively few degrees of freedom) or assume the presence of a few outliers.

Environmental concentrations and other environmental variables tend to have positive skewness. Therefore, environmental statistics texts often focus on positive skew distributions such as the log-normal, gamma, and Weibull. Discussions of distributions with nonnormal kurtosis are somewhat more scarce.

Skewness and kurtosis can be characterized using familiar formulae, based on 3rd and 4th centered moments. Alternative, outlier-resistant statistics can be based on quantiles (e.g., Helsel and Hirsch 1992; Hoaglin et al. 1983).

3.2.1.2 Parametric Distributions Useful for Environmental Risk Assessment

For the most part, the distributions that have been used in risk assessment are well-studied distributions discussed in probability texts, and used in stochastic modeling in many disciplines (this applies in particular to the log-normal, gamma, exponential, Weibull, and beta distributions). Useful overviews may be found in texts on environmental statistics and risk assessment such as Gilbert (1987), Ott (1995), and Vose (2000). Some additional types may be of interest in certain situations. In particular,

- Finite mixture distributions may be valuable when a distribution appears to result from mixing of somewhat distinct subpopulations, e.g., if there appear to be multiple modes in a distribution.
- Zero-modified distributions may be useful if concentration data contain more nondetections than can be accounted for by censoring at the level of detection.
- Transformations of the data may be used to extend the applicability of a particular standard distribution, in practice usually the normal distribution. For example, a log-normal random variable is a random variable that is normal after logarithmic transformation. Power transformations are also widely used, e.g., with Box–Cox transformations.
- Systems of distributions, such as the Pearson system (Pearson 1894) and the Johnson system (Johnson et al. 1994), can be used to select a distribution based on the skewness and kurtosis, as well as mean and variance. The Student t and logistic distributions are symmetric (like the normal distribution) but have heavier tails than the normal distribution.
- An example of a nonparametric distribution is the empirical distribution (assign probability $1/n$ to each of n values in a sample). A 2nd popular nonparametric approach involves smoothing the empirical distribution by the kernel density estimation approach. The approach is often used for graphical exploration of data (providing a smooth graph analogous to a histogram) but has also been suggested in a risk assessment context (flood prediction as reviewed by Lall 1995).

3.2.2   FITTING DISTRIBUTIONS TO DATA

In the problem of selecting a distribution for a 1D model of variation, there are 2 kinds of variables, namely, 1) the data, which we know; and 2) distribution parameters, which will be assigned values based on the data. Here we will often follow statistical terminology by using the term "estimation" (of parameters) instead of "fitting." In statistical terminology, the values assigned to distribution parameters are termed "estimates"; the expressions used to compute estimates are "estimators."

3.2.2.1   Setting Parameters Equal to Statistics, Method of Moments (MOM)

The most familiar estimation procedure is to assume that the population mean and variance are equal to the sample mean and variance. More generally, the method of moments (MOM) approach is to equate sample moments (mean, variance, skewness, and kurtosis) to the corresponding population. Software such as Crystal Ball (Oracle Corporation, Redwood Shores, CA) uses MOM to fit the gamma and beta distributions (see also Johnson et al. 1994). Use of higher moments is exemplified by fitting of the log Pearson III distribution using the sample skewness, a procedure widely used in hydrology to represent the distribution of flood magnitudes.

The general strategy of equating parameters to statistics is of course not restricted to moments. Reliance on sample percentiles (e.g., sample median) can lead to estimators that are not excessively sensitive to outliers. In general, to fit a distribution with k parameters, k parameters must be equated to distinct sample statistics.

3.2.2.2   Maximum Likelihood (ML)

Maximum likelihood (ML) is the approach most commonly used to fit a parametric distribution (Madgett 1998; Vose 2000). The idea is to choose the parameter values that maximize the probability of the data actually observed (for fitting discrete distributions) or the joint density of the data observed (for continuous distributions). Estimates or estimators based on the ML approach are termed maximum-likelihood estimates or estimators (MLEs).

3.2.2.3   Least Squares (LS) and Generalizations
(Weighted LS and Generalized LS)

Least squares (LS) estimation minimizes the sum of squared deviations, comparing observed values to values predicted by a curve with particular parameter values. Weighted LS (WLS) can take into account differences in the variances of residuals; generalized LS (GLS) can take into account covariances of residuals as well as differences in weights. Cases of LS estimation include the following:

- Species sensitivity distributions are sometimes fitted by minimizing the sum of squared deviations between the empirical cumulative distribution function (cdf) and the fitted cdf.
- A special case of LS is the computation of an arithmetic average (the arithmetic average is the single value that minimizes the sum of squared deviations for the data). A weighted arithmetic average is the WLS solution in that situation.
- In meta-analysis, weighted averages are often used in order to incorporate standard errors (SEs) from measurements of parameters from independent studies. The weight for a given estimate is set equal to $1/SE^2$.

3.2.2.4   Unbiased and Minimum-Variance Unbiased
Estimation, Particularly for Variances

Bias corrections are sometimes applied to MLEs (which often have some bias) or other estimates (as explained in the following section, [mean] bias occurs when the mean of the sampling distribution does not equal the parameter to be estimated). A simple bootstrap approach can be used to correct the bias of any estimate (Efron and Tibshirani 1993). A particularly important situation where it is not conventional to use the true MLE is in estimating the variance of a normal distribution. The conventional formula for the sample variance can be written as $s^2 = SSR/(n - 1)$ where SSR denotes the sum of squared residuals (observed values, minus mean value); s^2 is an unbiased estimator of the variance, whether the data are from a normal distribution or otherwise. If the data are from a normal distribution, the MLE is actually SSR/n, which will tend to underestimate the variance. For a normal distribution, s^2 can be characterized as a bias-adjusted MLE. Alternatively, s^2 can be characterized as the minimum-variance unbiased (MVU) estimator of the variance for a normal distribution (the estimator with minimum variance, among unbiased estimators).

In the formula for s^2 the denominator quantity $n - 1$ is termed the "degrees of freedom." Comparing this to the denominator for the MLE, the subtraction of 1 in the s^2 expression is viewed as accounting for the estimation of a single fixed-effect parameter (the mean of the distribution). It is conventional to apply the same general idea in the case of more complex normal-theory statistical models including nonlinear regression, multiple regression, variance components models, and mixed models (McCulloch and Searle 2001). For these models, the procedures that are conventional for estimating variances and covariances (in contrast to the MLEs) account for the number of estimated, fixed-effects parameters (e.g., grand mean or regression coefficients). This is considered to reduce the bias, but the estimators are not always strictly unbiased for all statistical models where the approach is applied.

The relationships among these procedures are complex, and the methods are not necessarily mutually exclusive. For example, MOM and WLS estimators will turn out to be ML estimators in some situations.

3.2.2.5   Random Effects, Empirical Bayes, and Shrinkage

Empirical Bayes methodology and other kinds of shrinkage estimation may be considered in situations where there is some, perhaps limited information for a situation of specific interest, but also a desire to give some weight to data from situations less representative. The term "shrinkage" expresses the idea that an estimate from the situation of specific interest is "shrunk" toward some prior estimate such as an estimate from less strictly representative situations. As yet the methods have seen little or no use for pesticide ecological risk assessment in regulatory contexts.

A model-based shrinkage approach can be based on a "random-effects" statistical model. Such models can take into account differences among subsets of the data representing different situations, by assuming higher order frequency distributions. Examples are mixed models and variance components models. Estimates of model parameters can be used to compute weights that can be used in shrinkage estimation. For a particular, important case (the linear mixed model) the estimates are the so-called "best linear unbiased predictors" (BLUPs) (Robinson 1991; Littell et al. 1996). Approaches of this general type have been adopted in a number of disciplines (Robinson 1991). One important example is the Kriging procedure used in spatial statistics. A prediction is made at a location for which a measurement is not available, weighting the available data according to distance from the point of interest.

3.2.3   Evaluating and Comparing Estimators

For a given distribution there may be more than 1 reasonable way to estimate the parameters. We might like to use the "optimal" approach. While different definitions of optimality can lead to different estimators, it seems useful to consider the profile of optimality properties of competing estimators. Many useful properties are documented in texts on mathematical statistics. However, simulation studies will sometimes be useful for comparing the performance of alternative procedures. We suggest that a good rule of thumb may be to use estimators that appeal to common sense, are expected to make good use of the information, and are reasonably easy to compute. With regard to "use of information" we note the theoretical sufficiency argument possessed by certain estimators, namely, maximum-likelihood estimators (in many familiar cases) and minimum-variance unbiased estimators.

3.2.3.1 Frequentist Criteria for Evaluating Estimators, the Sampling Distribution

In classical statistics, the most important type of criterion for judging estimators is a high probability that a parameter estimate will be close to the actual value of the parameter estimated. To implement the classical approach, it is necessary to quantify the "closeness" of an estimate to a parameter. One may rely on indices of absolute, relative, or squared error. Mean squared error (MSE) has often been used by statisticians, perhaps usually because of mathematical convenience. However, if estimators are evaluated using Monte Carlo simulation it is easy to use whatever criterion seems most reasonable in a given situation.

The classical, frequentist approach in statistics requires the concept of the "sampling distribution" of an estimator. In classical statistics, a data set is commonly treated as a random sample from a population. Of course, in some situations the data actually have been collected according to a probability-sampling scheme. Whether that is the case or not, processes generating the data will be subject to stochasticity and variation, which is a source of uncertainty in use of the data. Therefore, sampling concepts may be invoked in order to provide a model that accounts for the random processes, and that will lead to confidence intervals or standard errors. The "population" may or may not be conceived as a finite set of individuals. In some situations, such as when forecasting a future value, a continuous probability distribution plays the role of the population.

Parameter estimates are computed from more or less random samples, and therefore are also random. Thus, we associate with an estimator a particular distribution, its "sampling distribution." The sampling distribution of an estimator is the distribution that results from basing estimates on random samples. Sometimes the sampling distribution of an estimator can be derived analytically. For example, if the data are from a normal distribution then the sample mean also has a certain normal distribution. For the sample variance, the distribution is scalable to a chi-square distribution with appropriate degrees of freedom. In general, we can use Monte Carlo and bootstrap simulation to characterize any sampling distribution of interest. It is hoped that the sampling distribution of an estimator will be such that there is a high probability of a parameter estimate close to the actual value of the parameter.

3.2.3.2 Mean Squared Error (MSE) of Estimators, and Alternatives

In statistical literature, it is common to quantify the performance of an estimator using mean squared error (MSE). MSE is the average squared deviation of estimates from the actual parameter value (averaging over the sampling distribution). MSE is frequently the default index in evaluation of estimators, perhaps due to mathematical convenience. However, if we rely on Monte Carlo simulation to evaluate estimators we are free to use whatever index seems most meaningful for a particular situation. For example, we can evaluate the probability of a given relative error, expressed as a percentage of the parameter value, or we can evaluate the probability of an error exceeding some critical magnitude.

One often encounters a distinction between "precision" and "accuracy." Accuracy relates to systematic deviation between parameter estimates and actual parameter values; precision relates to the spread in the distribution of estimates. This terminology is not often used explicitly in the estimation theory literature, but the concepts are often implicit.

MSE reflects a combination of the accuracy and precision of an estimator. A convenient feature is that MSE can be decomposed into parts that correspond to accuracy and precision.

3.2.3.3   Statistical Bias and Parameter Invariance

In general, bias refers to a tendency for parameter estimates to deviate systematically from the true parameter value, based on some measure of the central tendency of the sampling distribution. In other words, bias is imperfect accuracy. In statistics, what is most often meant is "mean-unbiasedness." In this sense, an estimator is unbiased (UB) if the average value of estimates (averaging over the sampling distribution) is equal to the true value of the parameter. For example, the mean value of the sample mean (over the sampling distribution of the sample mean) equals the mean for the population. This chapter adheres to the statistical convention of using the term bias (without qualification) to mean mean-unbiasedness.

The criterion of mean-unbiasedness seems to be occasionally overemphasized. For example, the bias of an MLE may be mentioned in such a way as to suggest that it is an important drawback, without mention of other statistical performance criteria. Particularly for small samples, precision may be a more important consideration than bias, for purposes of an estimate that is likely to be close to the true value. It can happen that an attempt to correct bias results in lowered precision. An insistence that all estimators be UB would conflict with another valuable criterion, namely "parameter invariance" (Casella and Berger 1990). Consider the estimation of variance. As remarked in Sokal and Rohlf (1995), the familiar sample variance (usually denoted s^2) is UB for the population variance (σ^2). However, the sample standard deviation ($s = \sqrt{s^2}$) is not UB for the corresponding parameter σ. That unbiasedness cannot be eliminated for all transformations of a parameter simply results from the fact that the mean of a nonlinearly transformed variable does not generally equal the result of applying the transformation to the mean of the original variable. It seems that it would rarely be reasonable to argue that bias is important in one scale, and unimportant in any other scale.

The use of mean-unbiasedness is often mathematically convenient as a means to represent accuracy. However, it seems just as useful to define bias by comparing the median of the sampling distribution to the parameter, rather than comparing the mean, as suggested by various authors (e.g., Cox and Hinkley 1974; Lehmann and Casella 1998; Kendall et al. 1987, Volume 2). With a median-unbiased estimator there is equal probability for underestimation and overestimation. It is easy to evaluate median bias of an estimator using simulation. Usually where one encounters a complaint about the (mean) bias of an estimator, no attempt has been made to evaluate median bias.

3.2.3.4 Robustness and Outlier Resistance

There is often a particular concern for the effects of outliers or heavy-tailed distributions when using standard statistical techniques. To address this type of a situation, a parametric approach would be to use ML estimation assuming a heavy-tailed distribution (perhaps a Student t distribution with few degrees of freedom). However, simple ad hoc methods such as trimmed means may be useful. There is a large statistical literature on robust and outlier-resistant methods. (e.g., Hoaglin et al. 1983; Barnett and Lewis 1994).

As with many terms used in the section, the term "robust" is often used differently by statisticians, relative to use by other scientists. In statistical terminology, the term robust denotes that a procedure will perform well under different situations (not only if a single particular model is assumed to be true). Often the term refers to outlier resistance, particularly relative to methods that are optimal under normality assumptions.

3.2.3.5 Consistency

The statistical concept of "consistency" embodies the idea we can obtain as much accuracy and precision as desired by collecting enough data. Technically, there are different definitions of consistency recognized by mathematical statisticians. Consistency is one example of an "asymptotic property."

3.3 SOME PRACTICAL ASPECTS OF THE SELECTION OF UNIVARIATE DISTRIBUTIONS

3.3.1 DATA FOR CHARACTERIZING VARIATION

3.3.1.1 Evaluating Data Representativeness

Data used to describe variation are ideally representative of some population of risk assessment interest. Representativeness was a focus of an earlier workshop on selection of distributions (USEPA 1998). The role of problem formulation is emphasized. In case of representativeness issues, some adjustment of the data may be possible, perhaps based on a mechanistic or statistical model. Statistical random-effects models may be useful in situations where the model includes distributions among as well as within populations. However, simple approaches may be adequate, depending on the assessment tier, such as an attempt to characterize quantitatively the consequences of assuming the data to be representative.

3.3.1.2   Preliminary Data Exploration

The data should be explored using tabulations, summary statistics, and graphs, before any distributions are fitted. Exploratory data analysis can help us to uncover unanticipated aspects of the data as soon as possible and may suggest appropriate types of distributions. Graphs that may be particularly useful include box plots (or box-and-whisker plots), stem-and-leaf plots, histograms, and kernel density plots (Hoaglin et al. 1983; Helsel and Hirsch 1992). While distribution fitting is not emphasized in the exploratory step, it may be useful to assess the distribution with reference to some default distribution such as a normal or log-normal distribution, perhaps using a P–P plot.

Some remarks are in order regarding the use of skewness and kurtosis statistics as reported by statistical software:

- An accurate estimate of skewness or kurtosis requires a large sample size (e.g., Kendall et al. 1987, Volume 1).
- There may be variation among definitions of skewness and kurtosis statistics in various sources. For example, the reported skewness may or may not be adjusted to have a value of zero for a symmetric distribution.
- In situations where the log-normal distribution may be viewed as a default distribution (e.g., concentration measurements that tend to have positive skewness), it may be of interest to compute skewness for logarithms of the variable.

3.3.2   Selecting a Distribution Type

3.3.2.1   Development of Default Distributions

Pesticide regulation makes use of measurements of specific fate and effects properties, as specified in laws such as the US Federal Insecticides Fungicides and Rodenticides Act (FIFRA). Studies are conducted according to relatively standardized designs. Particularly in this type of situation, it seems reasonable to develop default distributions for particular variables, as measured in particular, standardized studies. Default assumptions may relate to default distribution types, or default distribution parameters such as a coefficient of variation, skewness, or kurtosis. Default distributions may be evaluated in comparative studies that draw from multiple literature sources. Databases of pesticide fate and effects properties, such as those maintained by the USEPA Office of Pesticide Programs, may be useful for such comparative analyses.

Default distributions can be evaluated at 2 levels: 1) in comparative studies, one may compare how often alternative distributions better describe the data, e.g., based on goodness-of-fit statistics, and 2) before applying a default distribution in a particular situation, one should evaluate whether the distribution agrees with whatever data are available for that situation.

It is good to keep in mind that there will be a certain rate of false positives, i.e., incorrect rejection of an appropriate distribution, and a certain rate of false negatives. It is sometimes suggested that one should routinely evaluate some set of distributions and always use the one that fits best according to some criterion. This will result in the use of different distribution families for a given variable, when the variable is evaluated on different occasions, particularly with small datasets.

3.3.2.2   Quantifying Support for a Distribution Type

Indices are needed that can be used to rank or select alternative distributions based on how well they agree with a sample of data. Such indices may be particularly useful for comparative analyses designed to select default distribution types. There are various possibilities for useful indices:

- Use the results of a goodness-of-fit test.
- Use the R^2 statistic comparing the fitted cdf to the cdf of the empirical distribution.
- ML estimation optimizes the likelihood function. Use the optimized value of the log-likelihood function.

There will be some tendency for distribution types with more adjustable parameters to fit the data better, just because of the greater flexibility. The number of fitted parameters can be taken into account when ranking or selecting distributions, by applying a penalty that accounts for the number of parameters estimated from the data. The most popular procedure of this type is the Akaike information criterion (AIC), a penalized log likelihood. In the case where a distribution with fewer parameters is obtained by fixing the values for some parameters, in a distribution with more parameters, the statistical significance of additional parameters can be evaluated using a likelihood-ratio test (for example, an exponential distribution is a special case of the gamma distribution and also a special case of the Weibull distribution, obtained by setting gamma or Weibull parameters equal to particular values). In such a test, the null hypothesis is the distribution with fewer fitted parameters, the alternative the distribution with more fitted parameters.

3.3.2.3   Parametric versus Empirical Distributions

When enough data are available, the need to assume a specific parametric distribution can be avoided by using the empirical distribution. The empirical distribution based on n observations is the distribution that assigns equal probability ($1/n$) to each observed value. A particular focus of a workshop on distribution selection (USEPA 1998) was "considerations for choosing between the use of parametric distribution functions ... and empirical distribution functions." That report of the workshop emphasizes case-specific criteria.

The cdf of the empirical distribution converges in probability to the true cdf, as n increases. However, in small samples the empirical distribution may have some features that we do not want to extrapolate to the population. The empirical distribution is discrete (with positive probability only for observed values), whereas the population distribution may be conceived as continuous. With n too small there may actually be substantial probability of real-world values outside the range observed in our data, e.g., some real-world exposures larger than the largest observed concentration.

An approach suggested in USEPA (1998) is to supplement the empirical distribution with an exponential tail (the "mixed exponential approach"). An approach not mentioned is to use a smoothed empirical distribution (a continuous nonparametric distribution). The most likely approach would be to use a kernel smoother, e.g., as sometimes used in flood prediction to provide a distribution for flood magnitudes (review in Lall 1995). These procedures have the effect of adding a continuous tail to the distribution, extending beyond the largest observed value.

When a parametric distribution is fitted, each datum contributes to the estimate of each parameter or percentile. Whether this is good or not depends on whether the distribution to be fitted is reasonable. If it is assumed that one can identify the "true" distribution, the data will be used in a way that is in some sense optimal. In the real world, where the best distribution is uncertain, it may happen that estimated frequencies for one tail of a distribution are sensitive to observations on the other tail, e.g., estimates of high concentration percentiles are sensitive to observed low concentrations.

3.3.3 FITTING A DISTRIBUTION OF A PARTICULAR TYPE

3.3.3.1 Choice of Estimation Procedure

Some workshop participants suggested that the choice of a fitting procedure (e.g., MOM or ML) is not likely to be the most critical decision in a risk assessment. For relatively standardized analyses, one may wish to use relatively refined methods, even if those are not the easiest to implement. This may reduce inconsistencies among analyses, caused by use of different fitting procedures. In any case, the computations can be automated by software development. For analyses that are not very standardized, it is understood that cost-effectiveness will be a consideration and that fitting techniques will often be selected based on convenience. In practice, a general-purpose software package such as Crystal Ball is likely to be used without much emphasis on distribution-fitting criteria. Using appropriate statistical and graphical methods, one can determine whether a fitted distribution adequately represents the data.

ML is the approach most commonly used to fit a distribution of a given type (Madgett 1998; Vose 2000). An advantage of ML estimation is that it is part of a broad statistical framework of likelihood-based statistical methodology, which provides statistical hypothesis tests (likelihood-ratio tests) and confidence intervals (Wald and profile likelihood intervals) as well as point estimates (Meeker and Escobar 1995). MLEs are invariant under parameter transformations (the MLE for some 1-to-1 function of a parameter is obtained by applying the function to the untransformed parameter). In most situations of interest to risk assessors, MLEs are consistent and sufficient (a distribution for which sufficient statistics fewer than n do not exist, MLEs or otherwise, is the Weibull distribution, which is not an exponential family). When MLEs are biased, the bias ordinarily disappears asymptotically (as data accumulate). ML may or may not require numerical optimization skills (for optimization of the likelihood function), depending on the distributional model.

An approach that is sometimes helpful, particularly for recent pesticide risk assessments, is to use the parameter values that result in best fit (in the sense of LS), comparing the fitted cdf to the cdf of the empirical distribution. In some cases, such as when fitting a log-normal distribution, formulae from linear regression can be used after transformations are applied to linearize the cdf. In other cases, the residual SS is minimized using numerical optimization, i.e., one uses nonlinear regression. This approach seems reasonable for point estimation. However, the statistical assumptions that would often be invoked to justify LS regression will not be met in this application. Therefore the use of any additional regression results (beyond the point estimates) is questionable. If there is a need to provide standard errors or confidence intervals for the estimates, bootstrap procedures are recommended.

In case of a need to compare alternative estimators, the preceding section provides information on criteria for evaluation of estimators. The performance of alternative estimators can be characterized using Monte Carlo simulation (e.g., Gilliom and Helsel 1986).

3.3.3.2 Possible Problems with Estimators That Rely on Logarithmic Transformation

When ML or MVU criteria are applied with the log-normal or gamma distributions, the computations involve logarithmic transformation of the data. In practice, it seems that effects of logarithmic transformation may be a particular concern when there are rounding problems or other complications associated with the smallest observations. In this type of situation, logarithmic transformation may be avoided by use of MOM estimation, with moments computed in the original scale. For the log-normal and gamma distributions it may be convenient to make use of the sample coefficient of variation, computed without logarithmic transformation (McCulloch and Nelder 1989, p 296; Millard and Neerchal 2000).

3.3.3.3 Correcting Parameter Estimates for Statistical Bias

Based on the discussion of criteria for parameter estimation, it is not necessarily important to use estimators that are "unbiased" in the statistical sense. The emphasis should be on the overall performance of the estimator, considering precision as well as accuracy. If bias is known to be large for practical purposes, bias correction may improve performance (bootstrap bias correction is easy). However, in practice, precision may be a greater concern than bias, particularly with few data, and bias correction may result in lower precision.

A particular situation where bias may be important is in statistical meta-analysis, where statistical estimates are combined across studies. When estimates from individual studies may be averaged arithmetically, it is better to average unbiased estimates (Rao 1973, Section 3a). In case of biases that are consistent across studies, an arithmetic average would have a bias of the same sign, regardless of the number of studies included in the analysis. The average of biased estimates could fail to be consistent (in the statistical sense).

3.3.3.4   Bounding a Distributed Variable

For certain distributions, the set of values for which the pdf is positive (the support) is unbounded. For example, the pdf of the log-normal distribution is positive for all positive real numbers. Ordinarily, there will be values too extreme to be reasonable, and so it is common to place bounds on the support. However, selecting precise values for the bounds may be a difficult decision.

Supposing that one has decided on bounds for a variable, one can fit a distribution that has a bounded support, such as the beta distribution or Johnson SB distribution. Alternatively, in a Monte Carlo implementation, one may sample the unbounded distribution and discard values that fall beyond the bounds. However, then a source of some discomfort is that the parameters of the distribution truncated in this way may deviate from the specification of the distribution (e.g., the mean and variance will be modified by truncation). It seems reasonable for Monte Carlo software to report the percentage discarded, and report means and variances of the distributions as truncated, for comparison to means and variances specified.

3.3.4   Assessing the Fit of a Distribution

A fitted distribution should be evaluated using graphical methods as well as statistical goodness-of-fit (GoF) tests. Appropriate procedures are available in texts on environmental statistics and risk assessment (e.g., Gilbert 1987; Helsel and Hirsch 1992; Millard and Neerchal 2000). It is suggested that USEPA (1998) be consulted regarding a number of practical considerations.

Some statistical tests are specific for evaluation of normality (log-normality, etc., normality of a transformed variable, etc.), while other tests are more broadly applicable. The most popular test of normality appears to be the Shapiro–Wilk test. Specialized tests of normality include outlier tests and tests for nonnormal skewness and nonnormal kurtosis. A chi-square test was formerly the conventional approach, but that approach may now be out of date.

According to USEPA (1998), "the group fully agreed that visualization/graphic representation of both data and the fitted distribution is the most appropriate and useful approach for ascertaining adequacy of fit. In general, the group agreed that conventional GoF tests have significant shortcomings and should not be the primary method for determining the adequacy of fit."

For graphical evaluation of distribution fit, probability (P–P) plots and quantile (Q–Q) plots are particularly helpful. Sometimes the R^2 statistic has been used to quantify the linearity of a P–P plot or Q–Q plot; however, in practice it appears that there may be substantial deviation between the observed and expected frequencies, despite an R^2 that would be viewed as large in many statistical contexts.

3.3.5 Detecting and Addressing Dependencies

Information on ways to handle dependencies and the consequences of ignoring dependencies are reviewed in risk assessment texts (e.g., Warren-Hicks and Moore 1998; Vose 2000).

3.3.5.1 Detecting Dependencies

Dependencies may be detected using statistical tests and graphical analysis. Scatter plots may be particularly helpful. Some software for statistical graphics will plot scatter plots for all pairs of variables in a data set in the form of a scatter-plot matrix. For tests of independence, nonparametric tests such as Kendall's τ are available, as well as tests based on the normal distribution. However, with limited data, there will be low power for tests of independence, so an assumption of independence should be scientifically plausible.

3.3.5.2 Dependent Actual Values or Dependent Statistical Errors?

When a distribution has multiple parameters estimated from the data, statistical errors associated with estimates of different parameters will not generally be independent (an important exception is that when sampling from a normal distribution, the sample mean and sample variance are independent). A familiar example of dependent errors in statistical estimation is the correlation of slope and intercept estimates in linear regression. The correlation is negative unless the independent variable is centered, in which case the correlation is zero. Similarly, in probit analysis one often observes a negative estimated covariance for the probit slope and intercept. This does not necessarily indicate a correlation of true slope and intercept across a population of chemicals or species, or of slope and LC50, in the real world (there has been some controversy regarding whether any "real" dependence is anticipated). A rigorous treatment of all sources of variation should be possible, making use of a random-effects model with different correlation structures at different levels of variation. Statistical software for such modeling is increasingly accessible.

3.3.5.3 Accommodating Dependencies

Once it is decided that an assumption of independence is not supported, there are a number of approaches for building dependencies into the probabilistic model (Monte Carlo per se does not assume independence):

- Members of the Chapter 3 work group particularly emphasized avoidance of implausible combinations of variables. Some conjectured that substantial benefit may be had just by excluding combinations of variables that are unreasonable, i.e., dependencies in the tails may be substantially more important for practical purposes than dependencies close to the center of a distribution.
- Parametric techniques based on the multivariate normal (MVN) distribution are particularly well developed. Parameters of the MVN distribution include a covariance or correlation for each pair of variables, as well as a mean and variance for each variable.

- Methods based on the MVN distribution have been used particularly for autocorrelated data, for example, in time series analysis and geostatistics. Autocorrelation occurs when the same variable is measured on different occasions or locations. It often happens that measurements taken close together are more highly correlated than measurements taken less close together. Environmental data often have some type of autocorrelation.
- A nonparametric approach can involve the use of synoptic data sets. In a synoptic data set, each unit is represented by a vector of measurements instead of a single measurement. For example, for synoptic data useful for pesticide fate, assessment could take the form of multiple physical–chemical measurements recorded for each of a sample of water bodies. The multivariate empirical distribution assigns equal probability ($1/n$) to each of n measurement vectors. Bootstrap evaluation of statistical error can involve sampling sets of n measurement vectors (with replacement). Dependencies are accounted for in such an approach because the variable combinations allowed are precisely those observed in the data, and correlations (or other dependency measures) are fixed equal to sample values.
- Capabilities are available in risk assessment software for inducing rank correlations among variables with arbitrary parametric distributions (Warren-Hicks and Moore 1998; Vose 2000). Also see Vose for a discussion of the envelope method for handling dependencies.

3.4   USING SCANTY AND FRAGMENTARY DATA

The data available may be too few, nonrepresentative, censored, or available only in summary form.

3.4.1   SOME STATISTICAL CONSEQUENCES OF SMALL SAMPLE SIZES

Possible effects of too-small n include the following:

- Statistical tests will have relatively low power. In particular, there will be low power for testing the fit of a parametric distribution.
- The probability of selecting the most appropriate parametric distribution from a set of candidate distributions will be comparatively low.
- Use of the empirical distribution may be problematic in particular because of a relatively high probability of encountering values in the real world that are beyond the range observed in our data.
- Certain methods associated with normality, such as the t interval for the mean, are equally valid at all sample sizes, so long as normality and other assumptions are accurate, and in most situations will improve in performance with increasing n. With small n one has low power for testing distributional assumptions.
- Parameter estimates will have high statistical error. In principal, this can be accounted for by use of 2D methods, which make use of a parameter uncertainty distribution. However, approximations of the sampling distribution may be relatively poor at small sample sizes.

- As a rule, accurate estimation of lower distribution moments will require fewer data than accurate estimation of higher moments (e.g., fewer data are needed for a decent estimate of the mean than for a decent estimate of kurtosis).
- There may be particular difficulties in characterizing the tails of distributions, because the distribution tails relate to relatively infrequent events.
- Bayesian analyses may be relatively more dependent on the prior and less dependent on the data.

3.4.2 USING RELATIVELY GENERIC INFORMATION

When the data for a situation of specific interest are inadequate, a common approach is to make some use of more generic information, including possibly information less representative of the situation of specific interest. For example, if the information for a specific pesticide is inadequate, then some features of an analysis may be based on information from a set of other pesticides, considered to be comparable. This can be appropriate particularly for an early-tier assessment, in which case the criteria may be designed to be protective, so far as we can judge.

With regard to relevant statistical methodologies, it is possible to define 2 situations, which can be termed a meta-analysis context and a shrinkage estimation context. Similar statistical models, in particular random-effects models, may be applicable in both situations. However, the results of such a model will be used somewhat differently.

In the first situation we hope to define a generic distribution based on information from multiple studies, and no study is treated as more representative than another, for the situations where the distribution will be used. Generic assumptions may relate to type of distribution or to distribution parameters (e.g., coefficient of variation, skewness, or kurtosis). An important case is the determination of multiplicative safety factor based on a generic coefficient of variation, and assuming log-normality.

Methods of statistical meta-analysis may be useful for combining information across studies. There are 2 principal varieties of meta-analytic estimation (Normand 1995). In a "fixed-effects" analysis the observed variation among estimates is attributable to the statistical error associated with the individual estimates. An important step is to compute a weighted average of unbiased estimates, where the weight for an estimate is computed by means of its standard error estimate. In a "random-effects" analysis one allows for additional variation, beyond statistical error, making use of a fitted random-effects model.

There is some USEPA precedent for use of statistical meta-analysis in a regulatory context, including the recent meta-analysis of organophosphate-related acetylcholinesterase inhibition data and meta-analysis of epidemiological studies on effects of 2nd hand tobacco smoke exposure. Warren-Hicks and Moore (1998) provide some discussion of the potential applicability of meta-analysis to ecological risk assessments.

In the 2nd situation there is a desire to give greater weight to some data that are considered representative for the situation of interest. However, particularly if those data are limited, there may be a desire to give some weight to less representative data.

An approach that is sometimes adopted in this type of situation is to rely on a "data trigger" such that too few data will result in use of generic distributional assumptions, while adequate data will result in use of only the more specific data. Alternatively, some average of specific and generic information may be used. For example, in hydrology the log Pearson III distribution is often used for flood magnitudes. A parameter of the distribution is the skewness. Because the skewness is subject to high statistical error, it is sometimes recommended to average the skewness for a specific locality with a skewness characteristic of a wider area. Rigorous approaches that make use of random-effects modeling results to compute weights for different subsets of the data are possible.

3.4.3 MAXIMUM ENTROPY AND OTHER REPRESENTATIONS OF IGNORANCE

A relatively flat distribution can represent a situation of relative uncertainty. For example, when one has only a maximum and minimum, the conventional default distribution is uniform between those values. The main difficulty in determining the flat distribution, to be used in a situation of relative ignorance, is that a distribution that is flat in one scale may be far from flat in another.

Suppose, for example, we are given that a half-life for a 1st-order dissipation curve will be between 1 and 2 days. An equally valid way to describe degradation in this context is with the degradation rate (= (ln 2)/half-life). Therefore the information available can be represented just as appropriately by saying that the degradation rate will be 0.347 to 0.693 per day. However, the 2 approaches for applying the uniform default distribution (for rates versus for half-lives) seem to be drastically different for practical purposes. For example, according to the assumption of uniformly distributed half-life, the probability is 0.25 that the half-life exceeds 1.75; according to the assumption of uniformly distributed rate the probability of the same event is 0.14.

The maximum-entropy (maxEnt) approach involves the use of a measure of the uncertainty in a distribution (Shannon–Weaver entropy). The idea is to choose the distribution type that has maximum uncertainty subject to specification of some features of the distribution such as the range or a few moments or percentiles. Warren-Hicks and Moore (1998) list maxEnt solutions for a number of situations. In particular when only a min and max is available the maxEnt solution is the uniform distribution. The solution when the information available is the mean and variance, and the min and max are infinite, is the normal distribution.

The maxEnt approach suffers from the scale dependence problem. Nevertheless, perhaps a distribution may be judged to be better than not assigning a distribution. Consequences of using uniform or maxEnt distributions for different scales can be explored in a sensitivity analysis. An additional difficulty is that in order to apply the maxEnt approach, particular features of a distribution may need to be assumed known when those features may actually be substantially uncertain.

Special uninformative distributions are often used in Bayesian analysis to represent prior parameter uncertainty, in cases of minimum prior information on the parameters. The idea is often to select a prior distribution such that the results of the analysis will be dominated by the data and minimally influenced by the prior.

3.4.4 JUDGMENT-BASED DISTRIBUTIONS AND BAYESIAN METHODOLOGY

The topic of eliciting probability distributions that are based purely on judgment (professional or otherwise) is discussed in texts on risk assessment (e.g., Moore 1983; Vose 2000) and decision theory or Bayesian methodology (e.g., Berger 1985). Elicitation methods may be considered with 1D models in case no data are available for fitting a model. In the 2D situation, elicitation may be used for the parameter uncertainty distributions. In that situation, it may happen that no kind of relative frequency data would be relevant, simply because the distributions represent subjective uncertainty and not relative frequency.

One sometimes encounters remarks on Bayesian methodology suggesting that the essence of the approach is a substitution of professional judgment for data, used in case the latter is substantially lacking. While this viewpoint contains a kernel of truth, the chapter on Bayesian methods provides a more complete picture of the approach (see Chapter 5 of this book). Bayesian methodology does provide tools for integration of information, possibly for very different types. Thus, the approach may be valuable for ensuring use of as much as possible of the (possibly limited) information available.

Elicitation of judgment may be involved in the selection of a prior distribution for Bayesian analysis. However, particularly because of developments in Bayesian computing, Bayesian modeling may be useful in data-rich situations. In those situations the priors may contain little prior information and may be chosen in such a way that the results will be dominated by the data rather than by the prior. The results may be acceptable from a frequentist viewpoint, if not actually identical to some frequentist results.

3.4.5 MEASUREMENTS ENTIRELY OR PARTIALLY MISSING

This category of difficulties includes the following:

- For some units, we know only that the value of interest falls beyond a certain value (right or left censoring).
- The range of the data takes the form of counts for "bins" associated with ranges of the variable (interval censoring).
- Units meeting certain conditions have been deleted from the data, without a record of how many units have been deleted (truncation).
- It is desirable to characterize dependencies by fitting a multivariate distribution. However, some measurements are missing for some units.

In general, these types of problems can be handled parametrically by formulating a stochastic model of how the data have been generated, including the mechanism of missingness (censoring, etc.), and then fitting the distribution using ML. Statistical error can be addressed using likelihood-based standard errors or confidence intervals. In addition to ML, various special methods may be applicable, depending on the context. For example, the presence of nondetections has no effect on percentiles above the detection limit(s), and so those may suffice for point estimates of model parameters.

It is common to suggest using ML and related procedures to model nondetections in concentration measurements. The missing-data mechanism is assumed to be that a nondetection occurs if and only if the value happens to fall below the detection limit, i.e., we assume left censoring (however, see Lambert et al. 1991). Methods are reviewed in texts on environmental statistics (e.g., Gilbert 1987; Helsel and Hirsch 1992; Gibbons 1994; Newman 1995; Millard and Neerchal 2000). The computer package UnCensor (Newman, Greene, and Dixon) is available on the World Wide Web.

Implementation of ML is straightforward in many cases. More difficult situations may involve a need to incorporate random effects, covariates, or autocorrelation. The likelihood function may involve difficult or intractable integrals. However, recent developments in statistical computing such as the EM algorithm and Gibbs sampler provide substantial flexibility for such cases (in complicated situations, a specialist in current statistical computing may be helpful). Alternatively, the GLS approach described below may be applicable.

3.4.6   USING SUMMARY STATISTICS, PARTICULARLY BY LS, WLS, AND GLS

In some cases the only information available may be a table of summary statistics. This type of situation is especially prevalent in regulatory contexts, where decisions may be made from data from different sources, summarized in different ways. The statistics available may be means and variances, confidence bounds, ranges, percentiles, and so on. The procedures that can be applied will depend on the statistics available and the distribution to be fitted, but it is possible to sketch a general approach. ML provides a relatively refined approach that may be practical in some situations. However, in many cases ML will be difficult because of the presence in the likelihood function of high-dimensional integrals. A more practical approach can be based on LS, WLS, or GLS; however, some skill with numerical methods and software will still be helpful. Assume availability of n_s summary statistics computed from n original data values. The statistics should be mathematically independent, that is, it should not generally be possible to compute one statistic precisely given values of the others, but they need to be statistically independent. These summaries will be used to assign values to p distribution parameters. We assume $p \leq n_s < n$. [If $n_s < p$ we do not have enough information to estimate all distribution parameters. If $n_s \geq n$, then we can reconstruct the raw data using n independent summary statistics, using a nonlinear equation solver if necessary.] If $n_s = p$ then the parameters are estimated by solving (using a nonlinear solver if necessary) a system of p equations in p unknowns. If $n_s > p$ then we can we have an "overspecified" problem. Assuming that each statistic has an obvious parametric analogue (as for means or quantiles) we may use LS to compute a set of parameter estimates close to the corresponding statistics. Refinements based on WLS or GLS can account for sampling correlations of the different statistics, or differences in the respective sampling variances. If necessary, variances and covariances of statistics may be based on some form of simulation. Once an estimation procedure has been devised, the sampling error of the estimate may be evaluated using parametric bootstrap.

3.5 REFERENCES

Balanda KP, MacGillivray HL. 1988. Kurtosis: a critical review. Am Stat 42(2):111–119.

Barnett V, Lewis T. 1994. Outliers in statistical data. New York: John Wiley & Sons.

Berger JO. 1985. Statistical decision theory and Bayesian analysis. New York: Springer.

Casella G, Berger RL. 1990. Statistical inference. Pacific Grove (CA): Duxbury.

Cox DR, Hinkley DV. 1974. Theoretical statistics. New York: Chapman and Hall.

Efron B, Tibshirani RJ. 1993. An introduction to the bootstrap. Monographs on Statistical and Probability, 57. New York: Chapman and Hall.

Gibbons RD. 1994. Statistical methods for groundwater monitoring. New York: Wiley.

Gilbert RO. 1987. Statistical methods for environmental pollution monitoring. New York: Van Nostrand Reinhold.

Gilliom RJ, Helsel DR. 1986. Estimation of distributional parameters for censored trace level water quality data: 1. Estimation techniques. Water Resour Res 22:135–146.

Helsel DR, Hirsch RM. 1992. Statistical methods in water resources research. Amsterdam (The Netherlands): Elsevier.

Hoaglin DC, Mosteller F, Tukey JW. 1983. Understanding robust and exploratory data analysis. New York: Wiley.

Johnson NL, Kotz S, Balakrishnan N. 1994 (vol 1), 1995 (vol 2). Continuous univariate distributions. 2nd ed. New York: Wiley.

Kendall M, Stuart A, Ord JK. 1987. Kendall's advanced theory of statistics (2 volumes). 5th ed. New York: Oxford University Press.

Lall U. 1995. Recent advances in nonparametric function estimation: Hydraulic applications. Rev Geophys 33 Suppl. Available from www.agu.org/revgeophys/lall01/lall01.html

Lambert D, Peterson B, Terpenning I. 1991. Nondetects, detection limits, and the probability of detection. J Am Stat Assoc 86(414):266–277.

Lehmann EL, Casella G. 1998. Theory of point estimates. 2nd ed. New York: Springer.

Littel RC, Milliken GA, Stroup WW, Wolfinger RD. 1996. SAS system for mixed models. Cary (NC): SAS Institute Inc.

Madgett A. 1998. Some uses for distribution-fitting software in teaching statistics. Am Stat 52(4):253–256.

McCulloch CE, Searle SR. 2001. Generalized, linear, and mixed models. New York: Wiley.

McCulloch P, Nelder JA. 1989. Generalized linear models. 2nd ed. New York: Chapman & Hall/CRC.

Meeker WQ, Escobar LA. 1995. Teaching about approximate confidence regions based on maximum likelihood estimation. Am Stat 49(1):48–52.

Millard SP, Neerchal NK. 2000. Environmental statistics with S-Plus. Boca Raton (FL): CRC Press.

Moore PG. 1983. The business of risk. Cambridge (UK): Cambridge University Press.

Newman MC. 1995. Quantitative methods in aquatic ecotoxicology. Chelsea (MI): Lewis Publishers.

Normand S-LT. 1995. Meta-analysis software: a comparative review. Am Stat 49(3):297–308.

Ott WR. 1990. A physical explanation of the lognormality of pollutant concentrations. J Air Waste Manag Assoc 40:1378–1383.

Ott WR. 1995. Environmental statistics and data analysis. Chelsea (MI): Lewis Publishers.

Rao CR. 1973. Linear statistical inference and its applications. New York: Wiley.

Robinson GK. 1991. That BLUP is a good thing — the estimation of random effects. Stat Sci 6:32–34.

Sokal RR, Rohlf FJ. 1995. Biometry. 3rd ed. New York: WH Freeman and Co.

[USEPA] US Environmental Protection Agency. 1998. Report of the Workshop on Selecting Input Distributions for Probabilistic Assessments. EPA/630/R-98/004. Available from: www.epa.gov/ncea/pdfs/input/input.pdf

Vose D. 2000. Risk analysis. A quantitative guide. New York: Wiley.

Warren-Hicks WJ, Moore RJ. 1998. Uncertainty analysis in ecological risk assessment. Pensacola (FL): SETAC.

4 Monte Carlo, Bayesian Monte Carlo, and First-Order Error Analysis

W. J. Warren-Hicks, S. Qian, J. Toll, D. L. Fischer,
E. Fite, W. G. Landis, M. Hamer, and E. P. Smith

4.1 INTRODUCTION

Monte Carlo is a deceptively simple method that has gained prominence in the eco-logical risk sciences in recent years. Its appeal is warranted because the method is a straightforward approach for generating probability distributions and conducting uncertainty analyses in all aspects of a typical risk assessment. Using Monte Carlo techniques, the analyst can perform the basic mathematical requirements of a proba-bilistic risk assessment, including propagating parameter uncertainty in exposure or effects models into prediction uncertainty, generating distributions of exposure and effects, and combining exposure and effects distributions into a joint distribution of risk. Monte Carlo is relatively easy to use, and several interactive and easy-to-use commercial software programs are available. Monte Carlo is certainly the most popular statistical method currently in use for probability-based risk assessments, and from a broad perspective, it is an excellent choice for most applications. Because of its flexibility and range of application, Monte Carlo is an excellent uncertainty analysis tool.

The underlying theory of Monte Carlo analysis is grounded in the long-run fre-quency interpretation of statistics. In this sense, Monte Carlo analysis is an inher-ently frequentist (i.e., classical statistics) concept. In Monte Carlo analysis, samples are drawn from a distribution (the sampling distribution of the random parameter) that represents the uncertainty of a random parameter, like an input to a multiparam-eter exposure model. As more and more samples are drawn, the mean of the samples is assumed to converge to the most likely value of the parameter (expected value). This convergence assumption is the basis for Monte Carlo theory and, in practice, is implemented by the repeated drawing of samples from the random parameter sam-pling distribution (see Figure 4.1).

Uncertainty analysis for multiparameter models may require assigning sam-pling distributions to many random parameters. In which case, a single value is drawn from each of the respective sampling distributions during each Monte Carlo iteration. After each random draw, the generated values of the random parameters

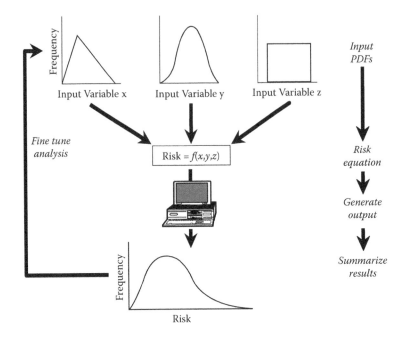

FIGURE 4.1   Monte Carlo analysis.

are plugged into the model, the model is run, and the model prediction(s) are collected. At the end of the Monte Carlo analysis, a histogram of the model predictions can be generated to view the output distribution (frequently termed the predictive distribution or Monte Carlo distribution) and descriptive statistics of the outputs can also be generated (median, mean, percentiles, range, and variance). The shape of the Monte Carlo output distribution can take on many forms and is directly dependent upon the choice of the input sampling distributions. In the case of a model-based uncertainty analysis, the Monte Carlo output distribution represents the uncertainty in model predictions, given the uncertainty in the model inputs. If the probability density function of the exposure distribution (generated, say, from a Monte Carlo analysis of uncertain model inputs) and effects distribution (generated, say, from a species sensitivity distribution) are available, Monte Carlo can be used to create a joint distribution of risk. The mathematical form of the sampling distribution(s) is not required, a simple column of values for each random parameter can be used as the basis for sampling (termed an empirical Monte Carlo analysis).

The popularity of Monte Carlo for risk-based uncertainty analysis is somewhat driven by the fact that Monte Carlo is fundamentally easy to implement, particularly with the advent of the personal computer, and graphically based software like Crystal Ball (www.decisioneering.com) and @Risk (www.palisade. com/risk.html). The availability of such software systems generally promotes the use of uncertainty analysis in ecological risk assessments, reducing the amount of mathematical and statistical knowledge required of the user to implement the method. While this development has mostly positive aspects, a danger arises for those users of the method that are not statistically trained. Statistically naive users are prone to making fundamental mathematical mistakes without the ability to judge the effect of such blunders. This issue is particularly relevant when choosing from the large number of available sampling distributions, use of advanced features such as 2nd-order Monte Carlo, or properly interpreting the output. Monte Carlo may be one of the most misunderstood uncertainty analysis approaches in common use, with many users captivated by the technique and graphics, disregarding the underlying theory and ramifications for interpretation once the analysis is complete.

In recent years, a large volume of literature has been produced on Monte Carlo analysis as an uncertainty method in ecological risk assessment (Warren-Hicks and Butcher 1996; Warren-Hicks and Moore 1998). This chapter does not attempt to review or summarize all of this information because so much of the current literature is inconsistent and the motivations for using the method are diverse. In the following narrative, we present the basics for the underlying theory of Monte Carlo and argue that informed users are better able to judge the validity of the Monte Carlo output than mathematically naive users. Monte Carlo is a practical and useful method, with an underlying statistical theory that has been proven over the years. In most standard ecological risk assessments, any major issue with Monte Carlo analysis is associated with the naïveté of the users of the technique, and not with any inherent limitations of the procedure.

4.2   PRACTICAL ASPECTS OF A MONTE CARLO ANALYSIS

As practiced, Monte Carlo is not only a statistical method, but also a process that involves numerous cascading decisions involving statisticians, toxicologists, and risk assessors. The degree of belief inherent in the Monte Carlo outputs is as much a function of the numerous decisions the investigator makes during the course of the analysis as it is the correct selection of the sampling distributions.

In most real-world problems, uncertainty is inherent in the choice of the analysis data set, treatment of outlying data points, choice of model, choice of spatial and temporal scales, choice of sampling distribution and associated parameters, etc. The analyst is faced with many decisions before implementing the Monte Carlo analysis and is subsequently faced with the challenge of interpreting the final output. Each choice the investigator makes plays a role in the interpretation of the Monte Carlo predictive distribution and in the expectation that decisions made based on the analysis are indeed correct. Because of the large number of decisions made by the investigator, the degree of belief that can be afforded the final prediction distribution that results from Monte Carlo is in large measure a function of the investigator's ability to implement the method and make appropriate decisions. In many cases, a poor Monte Carlo analysis can be attributed to the investigator's choices, rather than poor data quality. All investigators, after performing a Monte Carlo analysis, should ask: "Do I believe the answer?" "Does the shape of the distribution seem appropriate?" "Is the distribution skewed in the wrong direction?" "Is too much weight given to particular values of the predictions?" Because Monte Carlo is driven by the expertise of the investigator, all users of the software should look for ways to increase their degree of belief in the Monte Carlo result.

Monte Carlo sampling is discussed extensively in Hammersley and Morton (1956), Hammersley and Handscomb (1964), Kloek and Van Dijk (1978), and Wilson (1984). For Monte Carlo results to be believable, the convergence properties of the Monte Carlo estimators must be met. Several statistical and practical limitations exist in this regard. The most important practical limitations of Monte Carlo are the following:

1) Misspecification of the sampling distribution
2) Use of Monte Carlo sampling with a large number of assumed independent parameters, particularly when the parameters are highly correlated
3) Implementation with a relatively small number of iterations

For example, the distribution from which the samples are drawn is assumed to be the true distribution of the parameter of interest. To the degree that the sample distribution differs from the actual distribution (which is generally assumed unknown by the classical statistician), the confidence in the Monte Carlo results is decreased. Just how close these distributions must be is a complicated statistical issue that is frequently unclear. In a practical sense, if misspecification of a sampling distribution occurs for a very sensitive parameter in a multiparameter model, then the confidence in the Monte Carlo results is greatly diminished because the model prediction is greatly influenced by that parameter.

What is clear, however, is that the "garbage in, garbage out" adage applies. Two very important assumptions of Monte Carlo are the following: 1) the sampling distributions are the "true" distribution of the random parameter and 2) the Monte Carlo procedure is run to convergence. If the investigator uses incorrect sampling distributions (e.g., makes them up with little or no knowledge), the Monte Carlo results will effectively be incorrect. In addition, the underlying statistical theory behind Monte Carlo assumes that "enough" iterations are implemented for the convergence properties of the Monte Carlo estimators to hold. Again, the number of iterations required is not clear, particularly with disparate distributional assumptions among a large number of parameters. In hindsight, the investigator may actually have greater confidence in decisions based on a small number of measured data points in lieu of performing Monte Carlo analysis on a model for which basic parameterization and verification studies have not been implemented.

Burmaster and Anderson (1994) have proposed 14 "principles of good practice" for using Monte Carlo techniques. They suggest that before an analyst undertakes a Monte Carlo risk assessment, the growing literature on probabilistic risk assessment should be thoroughly examined. Principles for a properly conducted Monte Carlo analysis have also been proposed by the USEPA (1997).

4.3   MATHEMATICAL AND STATISTICAL UNDERPINNINGS OF MONTE CARLO METHODS

Many analysts do not understand the mathematics underlying the Monte Carlo method. While simple in concept, the underlying theory is somewhat complex. An understanding of the theory is important from the following perspectives:

1) The analyst is better able to judge the effect of decisions made during the course of the analysis.
2) The analyst is better able to explain and communicate the results of the Monte Carlo analysis and the statistical endpoints.
3) The analyst is better equipped to combine the Monte Carlo results with other analyses in a complex risk framework (e.g., combining exposure and effects distributions into a risk distribution).

The Monte Carlo method provides approximate solutions to a variety of mathematical problems by performing statistical sampling experiments on a computer. The modern Monte Carlo method originated during the development of atomic energy in the post–World War II era, when it was used to provide solutions to integral-differential equations. Later, the concept of using sampling experiments on a computer came to prevail in many scientific disciplines. Compared with other numerical methods, the Monte Carlo method is efficient with regard to computing time and easy to implement and understand. Using Monte Carlo methods for simulating the propagation of input errors through model predictions was initiated by O'Neill (1973) and McGrath and Irving (1973).

   The most common applications of the Monte Carlo method in numerical computation are for evaluating integrals. Monte Carlo methods can also be used in solving systems of equations. All instances of Monte Carlo simulation can be reduced to the evaluation of a definite integral like the following:

$$\mu = \int_a^b f(x)dx$$

(4.1)

Formally, suppose we have a random variable, x, which has measurements over the range a to b. Also, assume that the probability density function of x can be written as $p(x)$. In addition, assume a second function g, such that $g(x)\, p(x) = f(x)$. For example, $g(x)$ could represent a dose–response function on concentration and $p(x)$ is the probability density function of concentration. The expected value (which is the "most likely" value or the mean value) of $g(x)$ is μ

$$E\big(g(X)\big) = \int_a^b g(x)\,p(x)\,dx$$

$$= \int_a^b f(x)\,dx$$

$$= \mu$$

(4.2)

Notice that Equation (4.2) can be reduced to the same form as Equation (4.1). Estimating the expected value of $g(x)$ is a familiar statistical problem. A natural way of doing this is to take a random sample from x_i with distribution $p(x)$ and use the sample mean of $g(x_i)$ as an estimate of μ, that is,

Step 1. Draw random samples from $p(x)$: $x_i \sim p(x)$, for $i = 1, ..., n$

Step 2. Calculate the sample mean:

$$\hat{\mu} = \frac{1}{n}\sum_{i=1}^n g\big(x_i\big)$$

(4.3)

This estimate has a variance of

$$\mathrm{Var}(\hat{\mu}) = \frac{1}{n}\int_a^b \big(g(x) - \hat{\mu}\big)^2 dx$$

(4.4)

As a simple example, suppose that x is a random variable with a uniform density over the interval $[a, b]$ with $p(x) = 1/(b - a)$. As a result, $g(x) = (b - a)f(x)$. The integral is estimated by

$$\mu = (b - a)\,E(f(X))$$

(4.5)

The sample mean is calculated as

$$\hat{\mu} = \frac{(b-a)}{n}\sum f(x_i)$$

(4.6)

where x_i are values of a random sample of size n from a uniform distribution over (a, b). The estimate is unbiased, and the variance of the estimate is

$$\mathrm{Var}\big(\hat{\mu}\big) = \frac{b-a}{n}\int_a^b \big(f(x) - \hat{\mu}\big)^2 dx$$

(4.7)

The estimate of μ is based on a sample of simulated data; as a result, sampling error is always associated with the estimate. The law of large numbers states that the sample mean converges to the true mean in probability as the sample size increases:

$$\lim_{n \to \infty} \Pr\left(\left|\hat{\mu} - \mu\right| < \varepsilon\right) = 1 \tag{4.8}$$

In other words, a large sample size is necessary to reduce this sampling error.

In addition to increasing the sample size, reducing the sampling error can be done through efficient sampling. The Latin hypercube sampling is the most frequently used sampling technique for reducing Monte Carlo sampling error (Beckman and McKay 1987; Stein 1987; Tang 1993).

This method is designed to reduce sampling variance when sampling from several covariates. The technique uses a balanced or partially balanced fractional factorial design to sample, such that the sampling variance would be small at a given sample size. The Latin hypercube method was developed by McKay et al. (1979) for providing input to a computer experiment. Many researchers show that using Latin hypercube sampling can reduce the variance of the Monte Carlo estimator (Beckman and McKay 1987; Stein 1987; Tang 1993).

4.4 BAYESIAN MONTE CARLO ANALYSIS

Bayesian Monte Carlo analysis is a refinement of generalized sensitivity analysis, a modeling technique developed by Hornberger and Spear (1980) for their work on eutrophication modeling of Peel Inlet, Western Australia. In generalized sensitivity analysis (Figure 4.2), an investigator parameterizes a multisite model with data from sites that are "generically similar" to a particular site of interest, runs the model using site-specific inputs, and calibrates the model with site-specific output data. Calibration is based on an acceptance–rejection procedure that compares the site-specific output

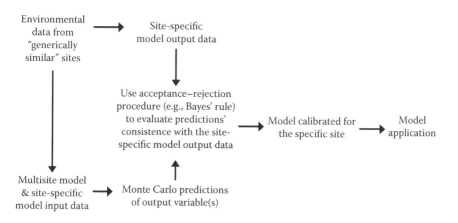

FIGURE 4.2 Generalized sensitivity analysis.

data to each realization of a Monte Carlo simulation and asks how well each of the model's predictions corresponds to the site-specific data. The generalized sensitivity analysis process uses Bayes's rule for the acceptance–rejection procedure

Hornberger and Spear's original application of generalized sensitivity analysis (GSA) used a binary acceptance–rejection procedure, i.e., they discarded a Monte Carlo realization if they thought that the prediction was inconsistent with the site-specific data (a "nonbehavior") or kept it if they thought it was consistent (a "behavior"). The prior probability on each Monte Carlo realization was the reciprocal of the total number of realizations. After the acceptance–rejection procedure was applied, the updated (posterior) probability on each realization that was classified as a behavior was the reciprocal of the number of behaviors, and the posterior probability on nonbehaviors was zero.

Variations on this simulation-based approach have been used to estimate model accuracy in water quality forecasts (Fedra 1983; Rose et al. 1991; van Straten and Keesman 1991), to compare chemical reaction model structures (Walter et al. 1986), and to estimate parameter values that produce a system response that satisfies given design criteria (Auslander et al. 1982). Related methods have descriptive names such as Monte Carlo filtering (Rose et al. 1991), Monte Carlo set-membership estimation (van Straten and Keesman 1991), range checking (Janse et al. 1992), feasible parameter space expansion (Li et al. 1994), and generalized likelihood uncertainty estimation (Beven and Binley 1992).

Although they did not make the link to generalized sensitivity analysis, Dilks et al. (1989, 1992) proposed using Bayes's rule to define the acceptance–rejection procedure for evaluating Monte Carlo simulation results. They coined the term Bayesian Monte Carlo analysis (BMC) for what amounted to doing generalized sensitivity analysis using Bayes's rule as the acceptance–rejection procedure. Subsequent applications of BMC include Patwardhan and Small (1992), Small and Escobar (1992), Brand and Small (1995), Dakins et al. (1996), Linkoff et al. (1999), and USEPA (2000). Other investigators were using Bayesian techniques in the late 1980s and early 1990s (e.g., Erdy 1989; Iman and Hora 1989; Wolpert et al. 1992), but the techniques were computationally limited until combined with Monte Carlo analysis.

By Bayes's rule, the posterior probability on a Monte Carlo realization of a model equals the probability of observing the site-specific output data if the realization is correct, times the prior probability that the realization is correct, normalized such that the sum of the posterior probabilities of the Monte Carlo realizations equals 1. In Monte Carlo analysis, all realizations are equally likely (i.e., the prior probability on each realization of an n-realization Monte Carlo simulation is $1/n$). Therefore, the BMC acceptance–rejection procedure boils down to the following: The probability that a model realization is correct, given new data, equals the relative likelihood of the having observed the new data if the realization is correct.

The posterior probabilities on the Monte Carlo realizations of the model are determined by the error structure in the data.

Consider, for example, a model that is being used to predict the log-concentration of a chemical in some environmental medium (e.g., the average log-transformed concentration in the muscle tissue of fish exposed to the chemical at a particular site).

A probabilistic model is available for predicting the average log-tissue residue as a function of the water concentration at the site, and a set of site-specific tissue residue measurements is available. The water concentration to which the fish were exposed is known, so the average log-tissue residue can be predicted with the model. A Monte Carlo simulation will provide a set of equally probable predictions of the average log-tissue residue. The BMC acceptance–rejection procedure then boils down to estimating, for each model prediction, the probability of getting the observed sample average log-tissue residue concentration if the model prediction is correct.

In this example, the likelihood function is the distribution on the average of a random sample of log-transformed tissue residue concentrations. One could assume that this likelihood function is normal, with standard deviation equal to the standard deviation of the log-transformed concentrations divided by the square root of the sample size. The likelihood function assumes that a given average log-tissue residue prediction is the true site-specific mean. The mathematical form of this likelihood function is

$$L\left(\log\left(M\hat{T}C_k\right)\right) = f\left[\left[\frac{1}{n_{obs}} \cdot \sum_{j=1}^{n_{obs}} \log\left(TC_{obs}\right)\right] \middle| \left(MTC = M\hat{T}C_k\right)\right]$$

$$= \frac{1}{\sqrt{2\pi s_{obs}^2/n_{obs}}} \cdot \exp\left[-\frac{1}{2} \cdot \left(\frac{\frac{1}{n_{obs}} \cdot \sum_{j=1}^{n_{obs}} \log\left(TC_{obs}\right) - \log\left(M\hat{T}C_k\right)}{\sqrt{s_{obs}^2/n_{obs}}}\right)\right] \quad (4.9)$$

where
$\log M\hat{T}C_k$ = log-transformed mean tissue residue concentration prediction
TC_{obs} = measured site-specific tissue residue concentration
s_{obs}^2 = sample variance of the log-transformed tissue concentration data
n_{obs} = number of site-specific tissue residue samples
j = index variable for site-specific tissue residue samples
k = index variable for MTC predictions

It answers the question "What is the probability of having obtained the observed site-specific sample average, if the prediction is the true site-specific mean?" The posterior probabilities on the Monte Carlo realizations of the model are given by

$$f''\left(M\hat{T}C_k\right) = \frac{f'\left(M\hat{T}C_k\right) \cdot L\left[\log\left(M\hat{T}C_k\right)\right]}{\sum_{k=1}^{n_{pred}} f'\left(M\hat{T}C_k\right) \cdot L\left[\log\left(M\hat{T}C_k\right)\right]} \quad (4.10)$$

The prior probability f' on each prediction ($1/n_{pred}$) cancels out, and the likelihood is given by Equation (4.9), leaving

$$f''\left(M\hat{T}C_k\right) = \frac{\dfrac{1}{\sqrt{2\pi s_{obs}^2/n_{obs}}} \cdot \exp\left[-\dfrac{1}{2} \cdot \left[\dfrac{\dfrac{1}{n_{obs}} \cdot \sum\limits_{j=1}^{n_{obs}} \log\left(TC_{obs}\right) - \log\left(M\hat{T}C_k\right)}{\sqrt{s_{obs}^2/n_{obs}}}\right]\right]}{\sum\limits_{k=1}^{n_{pred}} \dfrac{1}{\sqrt{2\pi s_{obs}^2/n_{obs}}} \cdot \exp\left[-\dfrac{1}{2} \cdot \left[\dfrac{\dfrac{1}{n_{obs}} \cdot \sum\limits_{j=1}^{n_{obs}} \log\left(TC_{obs}\right) - \log\left(M\hat{T}C_k\right)}{\sqrt{s_{obs}^2/n_{obs}}}\right]\right]} \quad (4.11)$$

Equation (4.11) is a useful form of the Bayesian acceptance–rejection procedure for generalized sensitivity analysis, in that it applies whenever one's model is predicting an average of a measured quantity.

Additional work on Bayesian Monte Carlo is found in Qian et al. (2003). This study examines the efficiency of Bayesian Monte Carlo techniques when a large number of unknown parameters are present in the model.

4.5   FIRST-ORDER ERROR ANALYSIS

First-order error analysis is a method for propagating uncertainty in the random parameters of a model into the model predictions using a fixed-form equation. This method is not a simulation like Monte Carlo but uses statistical theory to develop an equation that can easily be solved on a calculator. The method works well for linear models, but the accuracy of the method decreases as the model becomes more nonlinear. As a general rule, linear models that can be written down on a piece of paper work well with 1st-order error analysis. Complicated models that consist of a large number of pieced equations (like large exposure models) cannot be evaluated using 1st-order analysis. To use the technique, each partial differential equation of each random parameter with respect to the model must be solvable.

In the equation, Y is the model output, f is the model, and $(x_1, ..., x_p)$ are random model parameters with standard error $(S_1, ..., S_p)$. The variance of model output S_y^2 is given by the 1st-order Taylor expansion:

$$S_y^2 \approx \sum_{i=1}^{p} \left(\frac{\partial f}{\partial x_i}\right)^2 S_{x_i}^2 + 2 \times \sum_{j=1}^{p-1} \sum_{i=j+1}^{p} \left(\frac{\partial f}{\partial x_i}\right)\left(\frac{\partial f}{\partial x_j}\right) S_{x_i} S_{x_j} \rho_{x_i x_j} \quad (4.12)$$

where $\rho_{x_i x_j}$ is the correlation of x_i and x_j.

While the technique does require the analyst to remember simple calculus, the technique has numerous advantages over standard Monte Carlo approaches. The advantage of using the 1st-order error analysis is that we know the relative contribution of each uncertain variable to the uncertainty in the result. This relative contribution can be used to prioritize data collection efforts to reduce uncertainty in the parameter. In addition, we can easily see the effect of correlation on the model prediction variance. In the case where the model has 2 random parameters, notice that if $\rho_{x_i x_j}$ is negative (negative correlation), the resulting prediction error is smaller in the presence of correlation. If the correlation is positive, the resulting prediction error is larger in the presence of correlation. Another advantage of the method is that the exact sampling distribution is not required by the statistical theory. The variances are combined through the equation, regardless of the sampling distribution. Therefore, the investigator need not spend much time worrying about the exact sampling distribution. The investigator need only know the variance of the random parameter (in the proper units), for the underlying theory to hold.

As an example, in the following derivation the 1st-order error analysis equation for a simple model with both constants and random variables is found. The random terms are X and Z, with constants a, b, and c. The model is

$$Y = a + bX + cZ \qquad (4.13)$$

The 1st-order error analysis equation is then

$$\mathrm{Var}(Y) = \left(\frac{\partial Y}{\partial X}\right)^2 \sigma_X^2 + \left(\frac{\partial Y}{\partial Z}\right)^2 \sigma_Z^2 + 2\left(\frac{\partial Y}{\partial X}\right)\left(\frac{\partial Y}{\partial Z}\right)\sigma_X \sigma_Z \rho_{X,Z} \qquad (4.14)$$

The equation reduces to

$$\mathrm{Var}(Y) = b^2 \sigma_X^2 + c^2 \sigma_Z^2 + 2bc\rho_{X,Z} \qquad (4.15)$$

Values for the variance of X and Z, including the correlation of X and Z, can easily be plugged into Equation (4.15) and solved.

4.6 A MONTE CARLO CASE STUDY: DERIVATION OF CHRONIC RISK CURVES FOR ATRAZINE IN TENNESSEE PONDS USING MONTE CARLO ANALYSIS

Virtually no atrazine monitoring data exist for ponds. Many landscape factors affect atrazine concentrations in ponds, including the ratio of drainage area to water body volume, the proximity of treated fields to water, the percentage of crop area, tillage practices, soil property influences on atrazine degradation, and the geometry of the water bodies themselves. A Monte Carlo analysis was conducted to incorporate some of this variability into a pond water exposure model (Giddings et

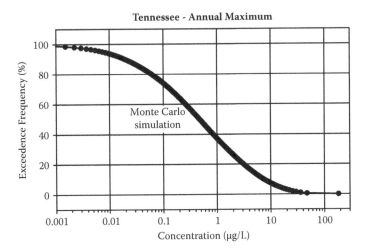

FIGURE 4.3   Exceedence curve for annual maximum atrazine concentrations in Tennessee pond water, based on exposure simulation using Monte Carlo analysis.

al. 2000, 2005). Model simulations linked the pesticide root zone model (PRZM), simulating surface runoff from treated fields; a runoff buffer model (RBUFF), simulating loss of atrazine between the edge of field and the water body; and a pond water quality model (PONDWQ) analogous to the exposure analysis modeling system (EXAMS) but able to simulate non–steady state hydrology. Scenarios were configured for a watershed in Tennessee for which field runoff data were available as benchmarks. The Monte Carlo analysis involved 14 000 simulations. Values for input parameters were selected from probability distribution functions. Model output for each simulation consisted of daily atrazine concentrations over a 36-year period, from which annual maximum values were determined. The distribution of the annual maximum concentrations (504 000 values) is shown as an exceedence curve in Figure 4.3.

Reliable chronic toxicity data were available for 21 species of plants (13 phytoplankton and 8 macrophytes) and 15 species of animals. The species sensitivity distributions (SSDs) for atrazine chronic toxicity (no observed effect concentrations [NOECs]) to plants and animals are shown in Figure 4.4. A log-normal distribution model was fitted to each SSD by least-squares regression.

The exposure distribution and species sensitivity distributions were integrated to generate risk curves for chronic effects. From the 504 000 values in the exposure exceedence curve, annual maximum concentrations corresponding to each 0.5th percentile were determined. The percentage of plant or animal species whose chronic NOEC would be exceeded at each of these concentrations was calculated from the log-normal SSD model. The percentage of plant or animal species affected at each exposure exceedence percentile was plotted as shown in Figure 4.5.

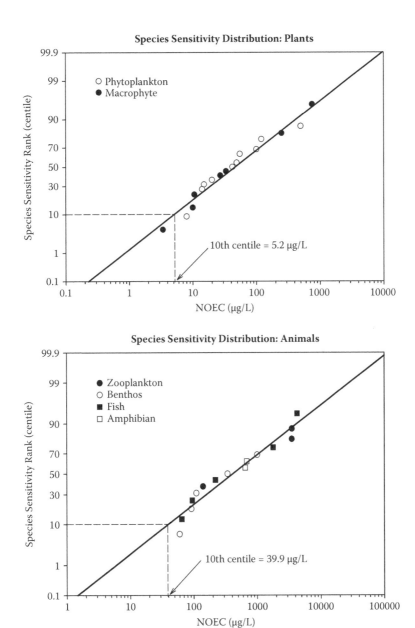

FIGURE 4.4 Species sensitivity distributions for chronic toxicity of atrazine to plants (upper panel) and animals (lower panel).

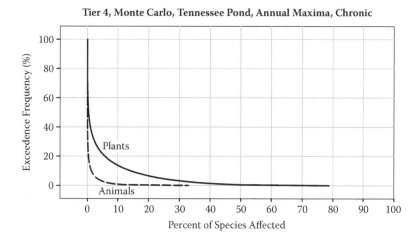

FIGURE 4.5 Risk curves based on exposure distribution for annual maximum atrazine concentrations in Tennessee ponds and chronic species sensitivity distributions for aquatic plants and animals.

4.7 CONCLUSIONS

Monte Carlo analysis, Bayesian Monte Carlo analysis, and 1st-order error analysis are useful tools for uncertainty analysis in risk assessment. Monte Carlo analysis may be the most flexible, user-friendly method available to risk assessors for error propagation. The method is useful for a wide range of applications, including error propagation, numerical evaluation of integrals, and combining exposure and effects distributions. First-order error analysis works well with simple models and has the distinct advantage of allowing the user to easily discern which random parameter is contributing most to the model prediction uncertainty. The method has a theoretical advantage in that the user is not required to predetermine the form of the sampling distribution for each random variate.

Based on our experience with Monte Carlo, we offer the following guiding principles for its use:

1) Use models that have been formally calibrated and validated for the site of interest.
2) Ensure the sampling distributions for uncertain model parameters are site specific and represent the actual quality and type of information available. For example, if only the possible ranges of a random parameter are available, do not select a distribution that places a large weight on the center of the parameter values. In this case, no information to justify the selected distribution is available. Distributions should reflect the available data. If no data or information about the distribution is available, consider not running Monte Carlo and finding alternative approaches for expressing uncertainty in the risk analysis.

3) Practice defending the chosen sampling distribution to a panel of experts. If you cannot think of quality reasons for the choice of distribution, do not use it.

4) Keep it simple. Generally, uncertainty analyses need not be complex, although many investigators tend to make them so. While available software enables rather complicated Monte Carlo approaches, simple approaches usually provide defendable results that can be easily communicated to others.

5) Determine whether or not actual measurements fall within the model prediction error. This may be the best way to justify the Monte Carlo result.

6) With simple linear models, try several uncertainty methods, like 1st-order error analysis, and judge whether the results are consistent.

7) Always perform a sensitivity test on the choice of sampling distribution and judge the degree to which the choice of distribution affects decisions made with the resulting Monte Carlo information.

8) If you can perform an uncertainty analysis with a calculator, do not use Monte Carlo analysis or at least compare the calculator and simulation method for consistency.

9) All choices should be questioned before the final analysis result is adopted. This includes data manipulation; methods for dealing with outliers; choice of sampling distribution, statistic, or model; implementation approaches; and number of samples in the simulation. These choices should be well documented in the description of the analysis.

In conclusion, we believe that error propagation methods like Monte Carlo, Bayesian Monte Carlo, and 1st-order error analysis should be promoted and extensively used in pesticide risk assessments implemented in both the United States and Europe.

4.8   REFERENCES

Auslander DM, Spear RC, Young GE. 1982. A simulation-based approach to the design of control systems with uncertain parameters. Trans Am Soc Mech Eng 104:20–26.

Beckman RJ, McKay MD. 1987. Monte Carlo estimation under different distributions using the same simulation. Technometrics 29:153–160.

Beven KJ, Binley AM. 1992. The future of distributed models: model calibration and uncertainty prediction. Hydrol Process 6:279–298.

Brand KP, Small MJ. 1995. Updating uncertainty in an integrated risk assessment: conceptual framework and methods. Risk Anal 15(6):719–731.

Burmaster DE, Anderson PD. 1994. Principles of good practice for the use of Monte Carlo techniques in human health and ecological risk assessments. Risk Anal 14:447–481.

Dakins ME, Toll JE, Small MJ, Brand K. 1996. Risk-based environmental remediation: Bayesian Monte Carlo analysis and the expected value of sample information. Risk Anal 16:67–69.

Dilks DW, Canale RP, Meier PG. 1989. Analysis of model uncertainty using Bayesian Monte Carlo. In: Proceedings of ASCE Specialty Conference on Environmental Engineering. New York: American Society of Civil Engineers, p 571–577.

Dilks DW, Canale RP, Meier PG. 1992. Development of Bayesian Monte Carlo techniques for water quality model uncertainty. Ecol Model 62:149–162.

Erdy DM. 1989. The confidence profile method: a Bayesian method for assessing health technologies. Operations Res 37:210–228.

Fedra K. 1983. A Monte Carlo approach to estimating and prediction. In: Beck MB, van Straten G, editors. Uncertainty and forecasting of water quality. Berlin: Springer-Verlag.

Giddings JM, Anderson TA, Hall LW, Hosmer, AR, Kendall RJ, Richards RP, Solomon KR, Williams WM. 2005. Atrazine in North America surface waters: a probabilistic aquatic ecological risk assessment. Pensacola (FL): SETAC.

Giddings JM, Anderson TA, Hall LW, Kendall RJ, Richards RP, Solomon KR, Williams WM. 2000. Aquatic ecological risk assessment of atrazine: a tiered, probabilistic approach. Novartis 709-00. Greensboro (NC): Novartis Crop Protection.

Hammersley JM, Handscomb DC. 1964. Monte Carlo methods. New York: Chapman and Hall.

Hammersley JM, Morton KW. 1956. A new Monte Carlo technique: antithetic variates. Proc Cambridge Philos Soc 52:449–475.

Hornberger GM, Spear RC. 1980. Eutrophication in Peel Inlet, I, problem-defining behavior and a mathematical model for the phosphorus scenario. Water Res 14:29–42.

Iman RL, Hora SC. 1989. Bayesian methods for modeling recovery time with an application to the loss of off-site power at nuclear power plants. Risk Anal 9:25–36.

Janse JH, Aldenberg T, Kramer PRG. 1992. A mathematical model of the phosphorus cycle in Lake Loosdrecht and simulation of additional measures. Hydrobiologia 233:119–136.

Kloek T, Van Dijk HK. 1978. Bayesian estimates of equation system parameters: an application of integration by Monte Carlo. Econometrics 46:1–20.

Li H, Watanabe K, Auslander D, Spear RC. 1994. Model parameter estimation and analysis: Understanding parametric structure. Ann Biomed Eng 22:97–111.

Linkoff I, Burmistrov D, Kandlikar M, Schell WR. 1999. Reducing uncertainty in the radionuclide transport modeling for the Chernobyl forests using Bayesian updating. In: Linkov I, Schell WR, editors. Contaminated forests. Dordrecht (DE): Kluwer, p 143–150.

McGrath EJ, Irving DC. 1973. Techniques for efficient Monte Carlo simulation. AD 762 721–723. Office of Naval Res. Resp. Springfield (VA): National Technological Information Service.

McKay MD, Conover WJ, Beckman RJ. 1979. A comparison of three methods for selecting values of input variables in the analysis of output from a computer code. Technometrics 21:239–245.

O'Neill RV. 1973. Error analysis of ecological models. In: Nelson DJ, editor, Conf. 710501, Radionuclides in ecosystems. Springfield (VA): National Technological Information Service.

Patwardhan A, Small MJ. 1992. Bayesian methods for model uncertainty analysis with applications to future sea level rise. Risk Anal 12:513–523.

Qian SS, Stow CA, Borsuk ME. 2003. On Monte Carlo methods for Bayesian inference. Ecol. Model 159:269–277.

Rose KA, Smith EP, Gardner RH, Brenkert AL, Bartell SM. 1991. Parameter sensitivities, Monte Carlo filtering, and model forecasting uncertainty. J Forecasting 10:117–133.

Small MJ, Escobar MD. 1992. Discussion of Paper by Wolpert et al. In: Gastonis C, Hodges XJ, Kass R, Singpurwalla N, editors. Bayesian statistics and technology: Case studies. New York: Springer-Verlag.

Stein M. 1987. Large sample properties of simulations using Latin hypercube sampling. Technometrics 29:143–151.

Tang B. 1993. Orthogonal array-based Latin hypercubes. J Am Stat Assoc 88:1392–1397.

[USEPA] US Environmental Protection Agency. 1997. Guiding principles for Monte Carlo analysis. EPA/630/R-97-001. Washington (DC): United States Environmental Protection Agency, Office of Research and Development.

[USEPA] US Environmental Protection Agency. 2000. Hudson River PCBs Reassessment RI/FS. Phase 2 Report, review copy. Further site characterization and analysis. Volume 2D, revised baseline modeling report. Book 3 of 4. Chapter 6. Prepared for USEPA Region 2 and US Army Corps of Engineers Kansas City District by TAMS Consultants, Limno-Tech Inc., Menzie-Cura and Associates Inc., and Tetra Tech Inc. January 2000.

van Straten G, Keesman KJ. 1991. Uncertainty propagation and speculation in projective forecasts of environmental change: a lake eutrophication example. J Forecasting 10:163–190.

Walter E, Piet-Lahannier H, Happel J. 1986. Estimation of non-uniquely identifiable parameters via exhaustive modeling and membership set theory. Math Computers Simulation 28:479–490.

Warren-Hicks WJ, Butcher B. 1996. Monte Carlo analysis: classical and Bayesian applications. Human Ecol Risk Assess 2:643–649.

Warren-Hicks WJ, Moore DRJ, editors. 1998. Uncertainty analysis in ecological risk assessment. Pensacola (FL): SETAC.

Wilson JR. 1984. Variance reduction techniques for digital simulation. Am J Math Manage Sci 4:277–312.

Wolpert RL, Steinberg LJ, Reckhow KH. 1992. Bayesian decision support using environmental transport and fate models. In: Gastonis C, Hodges XJ, Kass R, Singpurwalla N, editors. Bayesian statistics and technology: Case studies. New York: Springer-Verlag.

5 The Bayesian Vantage for Dealing with Uncertainty

D. A. Evans, M. C. Newman,
M. Lavine, J. S. Jaworska, J. Toll,
B. W. Brooks, and T. C. M. Brock

5.1 INTRODUCTION

Bayesian approaches are discussed throughout this book. Unfortunately, because frequentist methods are typically presented in introductory statistics courses, most environmental scientists do not clearly understand the basic premises of Bayesian methods. This lack of understanding could hamper appreciation for Bayesian approaches and delay the adaptation of these valuable methods for analyzing uncertainty in risk assessments.

Bayesian statistics are applicable to analyzing uncertainty in all phases of a risk assessment. Bayesian or probabilistic induction provides a quantitative way to estimate the plausibility of a proposed causality model (Howson and Urbach 1989), including the causal (conceptual) models central to chemical risk assessment (Newman and Evans 2002). Bayesian inductive methods quantify the plausibility of a conceptual model based on existing data and can accommodate a process of data augmentation (or pooling) until sufficient belief (or disbelief) has been accumulated about the proposed cause–effect model. Once a plausible conceptual model is defined, Bayesian methods can quantify uncertainties in parameter estimation or model predictions (predictive inferences). Relevant methods can be found in numerous textbooks, e.g., Carlin and Louis (2000) and Gelman et al. (1997).

Bayesian fundamentals are reviewed here because several chapters in this volume apply these methods in complex ways to assessing uncertainty. The goal is to create enough understanding so that methods described in later chapters can be fully appreciated.

5.2 CONVENTIONAL (FREQUENTIST) INFERENCE METHODS

The standard tools of statistical inference, including the concept and approaches of constructing a null hypotheses and associated p values, are based on the frequentist view of probability. From a frequentist perspective, the probability of an event is defined as the fraction of times that the event occurs in a very large number of trials (known as a probability limit). Given a hypothesis and data addressing it, the classical procedure is to calculate from the data an appropriate statistic, which is typically a single number. Based on the hypothesis being true and other assumptions, the probability distribution of this statistic is a known function.

This distribution, together with the numerical value of the statistic, allows an assessment of how "unusual" the data are, assuming that the hypothesis is valid. The p value is the probability that the observed value of the statistic (or values even more extreme) occur. The data are declared significant at a particular level (α); if $p < \alpha$, the data are considered sufficiently "unusual" relative to the hypothesis and the hypothesis is rejected. Standard, albeit arbitrary, values of α are taken as 0.05 and 0.01. Let us suppose that a particular data set gives $p = 0.02$. From the frequentist vantage, this means that, if the hypothesis were true and the whole experiment were to be repeated many times under identical conditions, in only 2% of such trials would the value of the statistic be "more unusual or extreme" than the value actually observed. One then prefers to believe that the data are not, in fact, "unusual"* and concludes that the assumed hypothesis is untenable.

It is important to note that the conclusion drawn from the observed data is based on a comparison with virtual data that might have been collected in other identical experiments but were never really observed. In fact, a judgement is made on the data rather than directly on the model or hypothesis. No consideration is given to the plausibility of the original hypothesis or specific alternatives. It is an erroneous assumption that the p value is a measure of the validity of the null hypothesis. As noted, p merely makes a statement about the data on the assumption that the hypothesis is valid.

While this is an almost universally used technique for testing hypotheses, the procedure can produce some odd or ambiguous conclusions. The following example, from the suggestion of Lindley and Phillips (1976), is quoted by Carlin and Louis (2000). We test the null hypothesis $H_0 : \theta = 0.5$ for the probability that a given coin will turn up "heads" after a toss; the alternative hypothesis is $H_a : \theta > 0.5$, i.e., the coin is biased toward "heads." Twelve independent tosses result in 9 heads and 3 tails. In this case, the choice of test statistic is simple; it is the number of heads, denoted by r. The binomial distribution gives the probability of obtaining r heads in 12 tosses as the following:

$$p\left(r \mid \theta, n = 12\right) = \binom{12}{r} \theta^r \left(1 - \theta\right)^{12-r} \tag{5.1}$$

The p value is given by

$$\sum_{r=9}^{r=12} p\left(r \mid \theta = 0.5, n = 12\right) = 0.073 \tag{5.2}$$

---

* That is, the observed statistic is considered to be a sample from the "center" of some (unknown) distribution whose form depends upon the true (unknown) hypothesis.

This is the probability of obtaining the observed number of heads, or more extreme (i.e., larger) values, when H_0 is assumed true ($\theta = 0.5$). Thus, H_0 is not rejected at the 5% level; to observe 9 heads, or more, in 12 tosses, is not sufficiently unusual for a coin with $\theta = 0.5$.

The above treatment has implicitly assumed that the experimental design was such that the number of trials was fixed at 12 and the observation was the number of heads. However, an alternative design could have been to continue tossing the coin until 3 tails were obtained, and the observation would be n, the number of tosses required to produce the 3 tails. In this case, the statistic for judging the data is just n. But the distribution of n, the number of tosses to produce 3 tails, is given by the negative binomial:

$$p\left(n\,|\,\theta, r = 3\right) = \binom{n-1}{2} \theta^{n-3}\left(1 - \theta\right)^3$$

(5.3)

and the p value for the experiment is given by

$$\sum_{n=12}^{\infty} p\left(n\,|\,\theta = 0.5, r = 3\right) = 0.033$$

(5.4)

This is the probability of a result of $n = 12$, or more extreme values, given that $\theta = 0.5$ (H_0 assumed true). The result calls for rejection of H_0 at the 5% level.

The difference arises because the identification of which of the data element is the random variable differs between the 2 designs. It is r, the number of heads, in the first case and n, the number of tosses, in the second. The p values compare the actually observed data with the data from an infinite number of virtual experiments (the frequentist approach). In the first case, all these experiments have 12 tosses and varying numbers of heads: in the second, they all have 3 tails and varying numbers of tosses.

Critics of the frequentist approach consider this disturbing. The actual observations: "in 12 tosses of a coin, 9 heads and 3 tails were observed" should not lead to 2 different conclusions dependent only upon the choice of when to stop the experiment (at 12 tosses or at 3 tails).

5.3 EXPERIMENTS CHANGE THE STATE OF KNOWLEDGE

The basic premise of the Bayesian approach is that observations change the state of knowledge of a system. Let us suppose for simplicity that the item of interest is some parameter, θ, describing a state of nature (as in the above example, where θ was a property of the coin and the conditions under which it was tossed). Figure 5.1 indicates symbolically the development of knowledge.

The extent of knowledge about θ can be quantified by showing that probability also can be interpreted as "degree of belief" (Lindley 1965), "measure of plausibility" (Loredo 1990), or "personal probability" (O'Hagan 2001). Early workers such as

FIGURE 5.1   Observations contribute to knowledge.

Bernoulli (1713) held this view of probability. Laplace (1812, 1820, 1951) described probability theory as "commonsense reduced to calculation." And, in Laplace's epistemic context, probability "expresses numerically degrees of uncertainty in light of data" (Howson and Urbach 1989). A large part of the motivation for the initial studies by workers such as Bernoulli and Laplace derived from the sponsorship of gambling noblemen. In fact, the problems addressed might now be called risk assessment because the noblemen wished to conduct their gaming so as to reduce their risk of loss and increase the "risk" of winning. Probability theory was intended to assist such decision making.

The above approach, which was attacked as being too vague to be the starting point of any theory of probability, led eventually to the frequentist approach, where probability was defined in a manner that assigns a numerical value, albeit a value that cannot ever be measured, since it requires an infinite number of trials

The numerical properties of probability and degrees of belief can be defined effectively and sensibly using a few axioms.

5.4   RULES OF PROBABILITY

Probability can be defined as a limiting case of a frequency ratio, and from this view the various rules of probability can be derived. An alternative approach is an axiomatic one that states that there is a quantity called probability associated with events and that it possesses assigned properties. The former is largely the frequentist point of view, the axiomatic approach is shared by Bayesians and non-Bayesians alike.

Probability values lie continuously in the range 0 to 1 inclusive, where the endpoints zero and unity are identified with impossibility and certainty, respectively. This follows immediately for the frequentist; for the axiomatic approach it is adopted as an axiom, but one imbued with Laplace's "commonsense." Any other range could be chosen at the cost of greater difficulty of interpretation.

Suppose A and B are events, then,

$$p(AB) = p(A|B) \cdot p(B) \tag{5.5}$$

where "AB" means that both events occur, or both propositions are valid. The notation introduces $p(A|B)$, the probability of A conditional on B. For the frequentist, it means $\lim_{N \to \infty} \left(n(AB)/n(B) \right)$, being the frequency ratio of occurrence of A on all the occasions when B occurred ($n \Rightarrow N$). If $p(B) \neq 0$ then, as the total number of trials (N) tends to infinity, so does $n(B)$. The above relationship then follows quite directly:

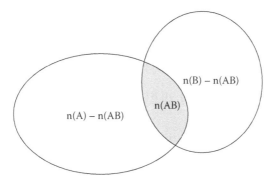

FIGURE 5.2 Venn diagram illustrating the development of conditional probability.

$$p\left(AB\right)=\lim_{N\to\infty}\left[\frac{n\left(AB\right)}{N}\right]=\lim_{N\to\infty}\left[\frac{n\left(AB\right)}{n\left(B\right)}\cdot\frac{n\left(B\right)}{N}\right]$$

$$=\lim_{N\to\infty}\left[\frac{n\left(AB\right)}{n\left(B\right)}\right]\cdot\lim_{N\to\infty}\left[\frac{n\left(B\right)}{N}\right]$$

$$=p\left(A\middle|B\right)\cdot p\left(B\right) \tag{5.6}$$

For the Bayesian, the relationship is taken as an axiom, but its motivation reflects the real world with the foreshadowing of rules implied by the above frequentist treatment.

Given the 2 events or propositions, A and B, then

$$p\left(A \text{ or } B\right)=p\left(A\right)+p\left(B\right)-p\left(AB\right) \tag{5.7}$$

where "A or B" means the inclusive "or," i.e., at least 1 of A and B occur. In Figure 5.2, it corresponds to the union of the 2 areas. The frequentist's numbers are shown in the various categories. From the figure it can be seen that

$$n\left(A \text{ or } B\right)=\left\{n\left(A\right)-n\left(AB\right)\right\}+\left\{n\left(B\right)-n\left(AB\right)\right\}+n\left(AB\right)$$

$$=n\left(A\right)+n\left(B\right)-n\left(AB\right) \tag{5.8}$$

from which the result follows. The Bayesian takes the result as an axiom, motivated by the real world.

5.5 BAYES' THEOREM

The result $p\left(AB\right)=p\left(A\middle|B\right)\cdot p\left(B\right)$ is symmetrical in A and B on the left side. It could equally well be written $p(BA)$, but

$$p(BA) = p(B|A) \cdot p(A) \qquad (5.9)$$

thus

$$p(A|B) \cdot p(B) = p(B|A) \cdot p(A)$$

$$p(A|B) = \frac{p(B|A) \cdot p(A)}{p(B)} \qquad (5.10)$$

Equation (5.10) is a statement of Bayes' theorem. Since the theorem is proved using results or axioms valid for both frequentist and Bayesian views, its use is not limited to Bayesian applications. Note that it relates 2 conditional probabilities where the events A and B are interchanged.

Bayesian interpretation and application of the theorem quantifies the development of information. Suppose that A is a statement or hypothesis, and let p(A) stand for the degree of belief in the statement or hypothesis A, based on prior knowledge, it is called the prior probability. Let B represent a set of observations, then p(B|A) is the probability that those observations occur given that A is true. This is called the "likelihood" of the data and is a function of the hypothesis. The left side, p(A|B), is the new degree of belief in A, taking into account the observations B, it is called the posterior probability. Thus Bayes' theorem tracks the effect that the observations have upon the changing knowledge about the hypothesis. The theorem can be expressed thus:

$$\text{posterior probability} \propto \text{likelihood function} \times \text{prior probability} \qquad (5.11)$$

Figure 5.3 is a copy of Figure 5.1 showing the portions of the Bayes formulation pertaining to each part of the development of knowledge.

The argument can be extended to treat multiple hypotheses. Suppose A_1 and A_2 are competing hypotheses. Then Bayes' theorem gives the following:

$$p(A_1|B) = \frac{p(B|A_1) \cdot p(A_1)}{p(B)}; \quad p(A_2|B) = \frac{p(B|A_2) \cdot p(A_2)}{p(B)} \qquad (5.12)$$

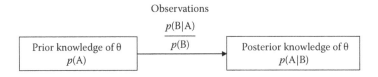

FIGURE 5.3   The contributions of the components of Bayes' theorem to the development of knowledge.

giving the ratio

$$\frac{p(A_1|B)}{p(A_2|B)} = \frac{p(B|A_1)}{p(B|A_2)} \cdot \frac{p(A_1)}{p(A_2)}$$

(5.13)

which can be written in words as

posterior odds ratio = likelihood ratio × prior odds ratio.

The coin-tossing experiment can be analyzed using this approach. As before, let θ be the probability of "heads." The hypothesis $\theta = 0.5$ is essentially meaningless because θ is a continuous parameter. Let the 2 hypotheses A_1 and A_2 be $\theta < 0.5$ and $\theta > 0.5$, respectively. The "prior odds ratio" represents an initial assessment of the relative probabilities or degree of belief of the 2 hypotheses. In the absence of any previous knowledge, a "noninformative prior" is used; in this case, we may assume it equally likely that the coin is biased to heads or tails, i.e., $p(A_1) = p(A_2)$. Consequently, the prior odds ratio is unity. Recognizing the earlier ambiguity whether the binomial or negative binomial distribution is applicable, we shall calculate the likelihood function, $p(B|A_i)$ using each.

For the binomial distribution the likelihood function is

$$p\left(9|n=12,\theta\right) = \binom{12}{9}\theta^9\left(1-\theta\right)^3$$

(5.14)

and for the negative binomial it is

$$p\left(12|\theta, r=3\right) = \binom{11}{2}\theta^9\left(1-\theta\right)^3$$

(5.15)

The functional dependence upon θ is identical for the 2 distributions. The coefficients cancel out when the likelihood is used in Bayes' theorem since they also appear in

$$p(B) = \int_0^1 p(B|\theta)d\theta$$

they also cancel out in the likelihood ratio. The approach does not suffer from ambiguity depending upon the design of the experiment; only the data are important in conformance with the likelihood principle, which states that the likelihood function expresses all the information that can be inferred about the parameter, θ, from the observed data.

Suppressing the unimportant coefficients, the likelihoods for the 2 hypotheses are obtained by integrating over the values of θ covering the appropriate range:

$$A_1, \quad p\left(r=9\big|n=12, \theta<0.5\right) \propto \int_0^{0.5} \theta^9 \left(1-\theta\right)^3 d\theta$$

$$= 1.613 \times 10^{-5}$$

$$A_2, \quad p\left(r=9\big|n=12, \theta>0.5\right) \propto \int_{0.5}^1 \theta^9 \left(1-\theta\right)^3 d\theta \qquad (5.16)$$

Thus the likelihood ratio is 0.048. The new state of knowledge concerning θ is then expressed by the posterior odds:

$$\frac{p\left(A_1\big|B\right)}{p\left(A_2\big|B\right)} = 0.048 \qquad (5.17)$$

This is a statement of the relative plausibility of the 2 hypotheses based on the observations. If one were a betting person, one would offer odds of 19 to 1 against the coin being biased toward "tails."

5.6 EXAMPLES RELEVANT TO UNCERTAINTY IN RISK ASSESSMENT QUANTIFYING PLAUSIBILITY OF A CAUSE–EFFECT MODEL

Central to any risk assessment is a model of causality. At the onset, a conceptual model is needed that identifies a plausible cause–effect relationship linking stressor exposure to some effect. Most ecological risk assessments rely heavily on weight-of-evidence or expert opinion methods to foster plausibility of the causal model. Unfortunately, such methods are prone to considerable error (Lane et al. 1987; Hutchinson and Lane 1989; Lane 1989), and attempts to quantify that error are rare. Although seldom used in risk assessment, Bayesian methods can explicitly quantify the plausibility of a causal model.

Let's use a fictitious example to illustrate the application of Bayes' theorem to quantifying the level of belief warranted in a causal model. A fishkill is observed below a discharge and the question is asked, "Did a toxic release (e.g., greater than LC10) from the point source cause the fishkill?" From the literature, one gathers evidence to assess this causal hypothesis. From a toxicological study of the major chemical of concern in the discharge, the likelihood of a fishkill if the discharge concentration was greater than LC10 (i.e., p(Fishkill|Release) > LC10) is calculated to be 0.400. From historical discharge records, it is also calculated that the probability of a discharge toxicant concentration greater than LC10 (i.e., p(Release) > LC10) is 0.005. From records of fishkills in this and similar streams of the region, the likelihood of a fishkill (p(Fishkill)) is 0.003. This information can be applied with Bayes' theorem (Equation (5.10)) to estimate the probability that there was a toxic discharge (>LC10) given the observed fishkill:

$$p(\text{Release} \mid \text{Fishkill}) = \frac{p(\text{Fishkill} \mid \text{Release}) \cdot p(\text{Release})}{p(\text{Fishkill})}$$

Inserting the above estimates into the right side of the equation gives a p(Release|Fishkill) of 0.666. Based on the evidence, the odds are 2 to 1 that the discharge caused the fishkill. Is this evidence sufficient to take regulatory action? Likely, it is not. One would need to gather more information in order to produce a level of belief sufficient to decide whether or not regulatory action was required. Assume that a characteristic lesion was found on the dead fish and that we know from the literature that p(Lesion|No Toxicant Exposure) = 0.010 and that p(Lesion|Exposure to the Discharge Toxicant) = 0.540. The likelihood ratio is 0.54:0.01 or 54:1. The posterior odds of 2:1 just calculated can become our new prior odds, and, based on this new evidence and Equation (5.11), new posterior odds of the toxic release having caused the fishkill can be calculated:

$$\text{Posterior Odds} = \text{Likelihood Ratio} \times \text{Prior Odds} = 54 \times 2 = 108{:}1$$

Based on this evidence, the odds that a toxic release caused the fishkill is a convincing 108 to 1. The level of belief is now sufficiently high for a reasonable person to take regulatory action. Bayes' theorem allowed optimal use of evidence to define the belief warranted in the causal hypothesis that a toxic release caused the fishkill: evidence changed our state of knowledge about the fishkill.

5.6.1 ESTIMATING INDOOR RADON EXPOSURE

Empirical Bayes methods were applied to estimate geometric means (GM) of indoor radon concentrations for Minnesota counties (Price et al. 1996). Data were collected unevenly among counties, with some counties having very low numbers of samples. Consequently, counties with low sample numbers had more error in GM estimates than adequately sampled counties. Bayesian methods allowed estimation of GM and associated variance despite these differences in county sample sizes. Even if no measurements were available for a given county, there is nonetheless some knowledge about the county GM. Denoting the logarithm of the GM by θ, the GM were assumed to be log-normally distributed among the counties based on existing data, i.e., the state of knowledge of θ is represented by $p(\theta) = N(\mu, \sigma^2)$, where μ = the "true" mean of the logarithm of radon concentration over all counties. The $p(\theta)$ is the informative prior distribution. Also, radon concentrations were judged to be log-normally distributed within counties based on results for amply sampled counties: $N(\mu, \sigma^2)$ for the logarithm of radon concentration. New estimates of county GM were then estimated with Bayes' theorem,

$$p(\theta|y) = \frac{p(\theta)\,p(y|\theta)}{p(y)}$$

(5.18)

where $p(\theta|y)$ = the probability that the true mean is θ given the data y, and $p(y|\theta)$ = the likelihood or probability of the data set, y, given θ. The $p(y)$ is a constant that, in practice, is estimated such that the right side of the equation integrates to unity (O'Hagan 2001). The "true" GM's of county radon concentrations were estimated with a modification of this equation and sample-size weighting of county geometric means. The informative prior distribution as modified by the likelihood of getting the data, y, for a county given θ and a better estimate of θ was produced: one "learned" from a particular county's data to produce a better estimate. The value of θ that maximizes Equation 5.18 can be considered a "best guess" of the true value of θ.

5.6.2 SPECIES SENSITIVITY EXAMPLE

Suppose that we wish to know the species sensitivity distribution (SSD) for a new pesticide, chemical A. Specifically, we wish to know the collection of LC50 values for many species. Unfortunately, chemical A has been tested on only a very limited number of species. For each species, an LC50 value has been estimated. Suppose also that pesticide B, having similar chemical structure and identical mode of action, has been tested on many species. Can we use the information about B to help us estimate the SSD for A and, if so, how? One way, of course, is informal. We take our knowledge of B and our subject matter knowledge, cogitate for a while, and come up with our best guess for the species sensitivity curve for A. However, a Bayesian approach provides a more formal, quantitative method for using the information about B.

We begin with a model for the shape of the SSD. For the sake of argument, we will assume that the SSD of B is approximately normal. That is, the histogram of the LC50 values for pesticide B looks approximately like a normal density with mean μ_B and variance σ_B^2. We may reasonably expect the SSD of A also to be normal with unknown mean μ_A but the same variance, $\sigma_A^2 = \sigma_B^2$. Standard statistical theory tells us how to estimate μ_A and σ_A^2 from the few species that have been tested with A. But Bayesian statistics goes a bit further by telling us also how to use the information about pesticide B.

The fact that A and B are so similar chemically suggests that their SSDs will also be similar. We can model that by saying that μ_A is likely to be within a range of μ_B plus or minus, for example, 200. This is formally expressed by a statement such as

$$\mu_A \sim N\left(\mu_B, 100^2\right)$$

(5.19)

i.e., μ_A has a normal distribution with mean μ_B and variance 100^2. The 100^2 arises because we treat the range of ± 200 as about 2 standard deviations; so 1 standard deviation is 100 and the variance is 100^2. Equation 5.19 is the prior probability distribution for μ_A. Suppose that each species tested with chemical A yields an LC50 value. Then Bayes' theorem and Bayesian statistics provide the formula for combining the prior distribution of μ_A with the data to yield the posterior distribution. Suppose that there were 4 species tested with A that yielded LC50 values of y_1, y_2, y_3, y_4.

The likelihood function, $p(\text{data}|\mu_A)$, for these data is

$$p\left(\{y_1, y_2, y_3, y_4\}|\mu_A\right) = p\left(y_1|\mu_A\right) \cdot p\left(y_2|\mu_A\right) \cdot p\left(y_3|\mu_A\right) \cdot p\left(y_4|\mu_A\right)$$

$$\propto e^{-(y_1-\mu_A)^2/2\sigma_B^2} \cdot e^{-(y_2-\mu_A)^2/2\sigma_B^2} \cdot e^{-(y_3-\mu_A)^2/2\sigma_B^2} \cdot e^{-(y_4-\mu_A)^2/2\sigma_B^2}$$

$$\propto e^{-\left[\sum(y_i-m_A)^2/\sigma_B^2\right]/2}$$

$$\propto e^{-\left[(\bar{y}-\mu_A)^2/(\sigma_B^2/4)\right]/2} \tag{5.20}$$

where $\bar{y} = (1/4)(y_1 + y_2 + y_3 + y_4)$, i.e., the mean of the observations. It is assumed that the observations are independent samples from a normal distribution with variance σ_B^2. The likelihood function of μ_A is a normal curve with the maximum at $\bar{y}$, and a variance of $\sigma_B^2/4$. Bayes' theorem gives the posterior distribution of μ_A as

$$p\left(\mu_A|\{y_1, y_2, y_3, y_4\}\right) = p\left(\{y_1, y_2, y_3, y_4\}|\mu_A\right) \cdot p\left(\mu_A\right)$$

$$\propto e^{-\left[(\bar{y}-\mu_A)^2/(\sigma_B^2/4)\right]/2} \cdot e^{-\left[(\mu_A-\mu_B)^2/100^2\right]/2}$$

$$\propto e^{-\left(\left[(\bar{y}-\mu_A)^2/(\sigma_B^2/4)\right]+\left[(\mu_A-\mu_B)^2/100^2\right]\right)/2} \tag{5.21}$$

Some algebra reveals that the posterior distribution of μ_A is normal with mean

$$\frac{\left(y_1 + y_2 + y_3 + y_4/\sigma_B^2\right) + \left(\mu_B/100^2\right)}{\left(4/\sigma_B^2\right) + \left(1/100^2\right)} \tag{5.22}$$

and variance

$$\frac{1}{\left(4/\sigma_B^2\right) + \left(1/100^2\right)} \tag{5.23}$$

These equations illustrate a common feature of Bayesian analysis: the posterior mean is a compromise between the prior mean and the data. In our example, as in every simple example with normally distributed data, the posterior mean is a weighted average of the prior mean and the data points. Each data point is weighted by the reciprocal of its variance, $1/\sigma_B^2$, just as the prior mean is weighted by the reciprocal of its variance, $1/100^2$. Because the reciprocal of a variance is such a useful concept, it is given a special name, precision. The posterior mean is just the weighted average of the prior mean and the data; the weights are the precisions. The general formula is the following:

$$\text{posterior mean} = \frac{\sum\left(\text{data precision} \times \text{data value}\right) + \text{prior precision} \times \text{prior mean}}{\sum\left(\text{data precision}\right) + \text{prior precision}}$$

$$\text{posterior precision} = N \times \text{data precision} + \text{prior precision}$$

$$\text{posterior variance} = \frac{1}{\text{posterior precision}} \tag{5.24}$$

Terms in the formula get more or less weight according to their precision, i.e., according to how accurate they are as measures of μ_A. The posterior precision measures how accurately we know μ_A. It is the sum of the prior precision and the precisions of each of the data points. In our example that is

$$\frac{1}{100^2} + \frac{1}{\sigma_B^2} + \frac{1}{\sigma_B^2} + \frac{1}{\sigma_B^2} + \frac{1}{\sigma_B^2} \tag{5.25}$$

This simple example illustrates principles of Bayesian analysis and how it accommodates information from different sources. Real situations and real analyses can be more complicated than our example. For example, when species are tested with chemical A, we might not know their LC50 values exactly; instead, we might have estimates of LC50 values. Or we may have data on another similar chemical C. In each case, we would adjust the analysis to accommodate the more complicated situation.

5.6.3 INFERENCE ABOUT CONFIDENCE INTERVALS

Confidence intervals are interpreted differently by frequentists and Bayesians. The 95% confidence interval derived by a frequentist suggests that the "true" value of some parameter (θ) will be contained within the interval 95% of the time in an infinite number of trials. Note that each trial results in a different interval because the data are different. This statement is dependent on the assumed conditions under which the calculations were done, e.g., an infinite number of trials and identical conditions for each trial (O'Hagan 2001). Nothing can be said about whether or not the interval contains the "true" θ.

The Bayesian approach reverses the role of the sample and model: the sample is fixed and unique, and the model itself is uncertain. This viewpoint corresponds more closely to the practical situation facing the individual researcher: there is only 1 sample, and there are doubts either what model to use, or, for a specified model, what parameter values to assign. The model uncertainty is addressed by considering that the model parameters are distributed. In other words Bayesian interpretation of a confidence interval is that it indicates the level of belief warranted by the data: the posterior probability is 0.95 that the "true" θ is within the stated 95% confidence interval. Statements are made about θ based on the data alone, not an infinite number of virtual trials.

The classical or frequentist approach to probability is the one most taught in university courses. That may change, however, because the Bayesian approach is the more easily understood statistical philosophy, both conceptually as well as numerically. Many scientists have difficulty in articulating correctly the meaning of a confidence interval within the "classical" frequentist framework. The common misinterpretation: the probability that a parameter lies between certain limits is exactly the correct one from the Bayesian standpoint.

Apart from this pedagogical aspect (cf. Lee 1989, preface), there is a more technical reason to prefer the Bayesian approach to the confidence approach. The Bayesian approach is the more powerful one eventually, for extending a model into directions necessary to deal with its weaknesses. These are various relaxations of distributional assumptions. The conceptual device of an infinite repetition of samples, as in the frequentist viewpoint, does not yield enough power to accomplish these extensions.

Confidence intervals using frequentist and Bayesian approaches have been compared for the normal distribution with mean μ and standard deviation σ (Aldenberg and Jaworska 2000). In particular, data on species sensitivity to a toxicant was fitted to a normal distribution to form the species sensitivity distribution (SSD). Fraction affected (*FA*) and the hazardous concentration (*HC*), i.e., percentiles and their confidence intervals, were analyzed. Lower and upper confidence limits were developed from t statistics to form 90% 2-sided classical confidence intervals. Bayesian treatment of the uncertainty of μ and σ of a presupposed normal distribution followed the approach of Box and Tiao (1973, chapter 2, section 2.4). Noninformative prior distributions for the parameters μ and σ specify the initial state of knowledge. These were constant c and $1/\sigma$, respectively. Bayes' theorem transforms the prior into the posterior distribution by the multiplication of the classic likelihood function of the data and the joint prior distribution of the parameters, in this case μ and σ (Figure 5.4).

The Bayesian equivalent to the frequentist 90% confidence interval is delineated by the 5th and 95th percentiles of the posterior distribution. Bayesian confidence intervals for SSD (Figures 5.4 to 5.5), 5th percentile, i.e., HC5 and fraction affected (Figures 5.4 to 5.6) were calculated from the posterior distribution. Thus, the uncertainties of both *HC* and *FA* are established in 1 consistent mathematical framework: *FA* estimates at the $\log_{10}$ HC lead to the intended protection percentage, i.e., $FA^{50}(\log_{10} HC_p^{50}) = p$ where p is a protection level. Further full distribution of *HC* and *FA* uncertainty can be very easily extracted from posterior distribution for any level of protection and visualized (Figures 5.5 to 5.7).

For the normal distribution there are analytical solutions allowing the assessment of both *FA* and *HC* using frequentist statistics. In contrast, Bayesian solutions are numerical. This highlights the flexibility of the Bayesian approach since it can easily deal with any distribution, which is not always possible with the frequentist approach.

Aldenberg and Jaworska (2000) demonstrate that frequentist statistics and the Bayesian approach with noninformative prior results in identical confidence intervals for the normal distribution. Generally speaking, this is more the exception than the rule.

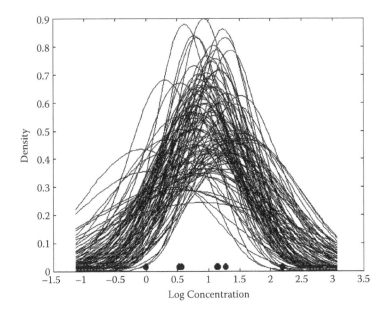

FIGURE 5.4 Bayesian normal density "spaghetti plot": random sample of 100 normal probability density functions (pdfs) drawn from the posterior distribution of μ and σ, given 7 cadmium NOEC toxicity data (dots) from Aldenberg and Jaworska (2000).

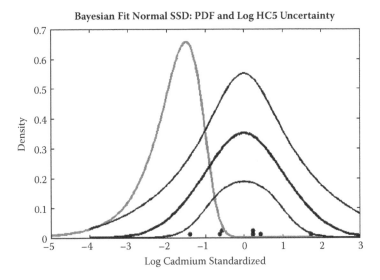

FIGURE 5.5 Bayesian posterior normal probability density function values for SSD for cadmium and its Bayesian confidence limits: 5th, 50th, and 95th percentiles (black) and Bayesian posterior probability density of the HC5 (gray).

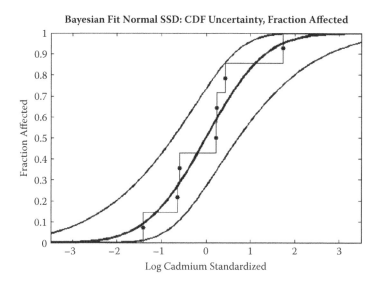

FIGURE 5.6 Bayesian confidence limits of the fraction affected: percentiles (5th, 50th, and 95th) of posterior normal cdfs for cadmium. Data plotted cumulatively at $(i - 0.5)/n$, with i rank order, and n the number of species tested.

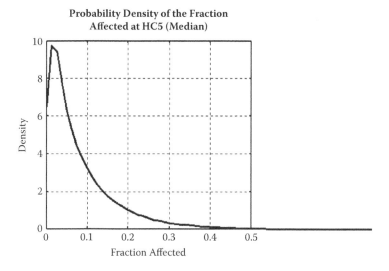

FIGURE 5.7 Bayesian posterior probability density of the fraction affected at median log (HC5) for cadmium.

For those who feel more confident with the frequentist approach and find the Bayesian approach controversial to some extent, it is advantageous that both approaches yield the same answers in this simplest case. This might add confidence in the Bayesian approach for some practitioners.

5.7   CONCLUSION

The general Bayesian context is presented in this brief chapter with the intent of building sufficient understanding so that the reader can fully appreciate the methods presented with more complexity in following chapters. The distinction between the frequentist and Bayesian vantages was made using contrasting analyses of the outcomes of simple coin-toss trials. Then, the Bayesian theorem and associated concepts were explored briefly. Three examples relevant to uncertainty in risk assessments were given: estimation of the level of belief warranted for a causal model, estimation of exposure concentrations based on uneven sampling of a study area, and interpretation of confidence intervals. Hopefully, more involved Bayesian methods applied in later chapters will now be more easily understood.

5.8   REFERENCES

Aldenberg T, Jaworska J. 2000 Uncertainty of the hazardous concentration and fraction affected for normal species sensitivity distributions. Ecotox Env Saf 46:1–18.

Bernoulli J. 1713. Ars Conjectandi. Basel.

Box GEP, Tiao GC. 1973. Bayesian inference in statistical analysis. New York: Wiley Classics.

Carlin BP, Louis TA. 2000. Bayes and empirical Bayes methods for data analysis. 2nd ed. Boca Raton (FL): Chapman and Hall/CRC.

Gelman A, Carlin JB, Stern HS, Rubin DB. 1997. Bayesian data analysis. Boca Raton (FL): Chapman and Hall/CRC.

Howson C, Urbach P. 1989. Scientific reasoning. The Bayesian approach. La Salle (IL): Open Court.

Hutchinson TA, Lane DA. 1989. Assessing methods for causality assessment of suspected adverse drug reactions. J Clin Epidemiol 42:5–16.

Lane DA. 1989. Subjective probability and causality assessment. Appl Stochastic Model Data Anal 5:53–76.

Lane DA, Kramer MS, Hutchinson TA, Jones JK, Naranjo C. 1987. The causality assessment of adverse drug reactions using a Bayesian approach. Pharm Med 2:265–283.

Laplace PS. 1812. Theorie analytique des proabilitiés. Paris: Courcier.

Laplace PS. 1820. Essai Philosophique sur les Probabilitiés.

Laplace PS. 1951. *Philosophical essay on probability*. New York: Dover Publications (originally published as the introduction to Laplace 1812, Paris: Courcier).

Lee PM. 1989. Bayesian statistics: An introduction. Oxford (UK): Oxford University Press.

Lindley DV. 1965. Introduction to probability and statistics from a Bayesian viewpoint. 2 vols. Cambridge (UK): Cambridge University Press.

Lindley DV, Phillips LD. 1976. Inference for a Bernoulli process (a Bayesian view). Am Stat 30:112–119.

Loredo TJ. 1990. From Laplace to Supernova SN 1987A: A Bayesian inference in astrophysics. In: Fougère PF, editor. Maximum entropy and Bayesian methods. Dordrecht (DE): Kluwer.

Newman MC, Evans DA. 2002. Causal inference in risk assessments: cognitive idols or Bayesian theory? In: Newman MC, Roberts M, Hale R, editors. Coastal and estuarine risk assessment. Boca Raton (FL): CRC Press, p 73–96.

O'Hagan A. 2001. Uncertainty in toxicological predictions: the Bayesian approach to statistics. In: Rainbow PS, Hopkin SP, Crane M, editors. Forecasting the environmental fate and effects of chemicals. Chichester (UK): John Wiley, p 25–41.

Price PN, Nero AV, Gelman A. 1996. Bayesian prediction of mean indoor radon concentrations for Minnesota counties. Health Phys 71:922–936.

6 Bounding Uncertainty Analyses

S. Ferson, D. R. J. Moore, P. J. Van den Brink,
T. L. Estes, K. Gallagher, R. O'Connor,
and F. Verdonck

6.1 INTRODUCTION

Risk analysts emphasize the differences between variability and incertitude, which are fundamentally different kinds of uncertainty. Variability is heterogeneity and stochasticity, such as spatial variation in chemical concentration, temporal fluctuations in weather, and genetic differences in susceptibility among individuals. Incertitude, on the other hand, is incomplete knowledge such as that arising from measurement error, doubt about the model or abstraction that should be used, limited sample sizes, possible biases in empirical design, and use of surrogate data. Incertitude can generally be reduced by additional empirical effort, but this is not true for variability. Although variability can perhaps be better characterized by the collection of more data, its amount and patterns are usually objective facts of nature that are not diminished by empirical effort. Most analysts agree that it is essential to keep these 2 kinds of uncertainty separate in any assessment for the sake of planning effective remediation or management strategies.

6.1.1 WHY BOUNDING?

There are basically 2 approaches for making calculations in the face of uncertainty. One approach is to approximate an estimate. Much of the machinery of modern analysis and statistics is focused on getting good approximations. Another approach is to bound the value being sought. These 2 approaches are clearly complementary to each other.

Rowe (1988) reviewed the following advantages of bounding over approximation as a strategy for calculation under uncertainty. Bounds are

- Possible even when estimates are impossible
- Rigorous rather than approximate
- Possible to make optimally tight
- Usually easier to compute than approximations
- Very simple to combine with other bounds
- Often sufficient for decisions

It is generally possible to obtain bounds on a quantity even when reasonable approximations are impossible. For some variable, we may have no reasonable basis to say what its value is in a particular situation, yet be entirely confident of bounds on it. Averaging upper and lower bounds to make an estimate destroys this confidence in order to construct an approximation of unknown reliability. If handled consistently, bounding can yield rigorous mathematical results, rather than mere approximations. Such calculations allow us to be sure, which can be more useful in practical settings than being, say, 95% sure, as statistical confidence procedures permit us to be. In many cases, bounding estimates can be shown to be the best possible. In other words, it may be possible to prove that the bounds could not be any tighter given the available information. Such a finding would give an analyst an excellent argument for gathering more data if the bounds were too wide to support clear decisions.

Bounds are often easier to compute than approximate estimates, which, in contrast, commonly require the solving of integrals. This simplicity of calculation extends to the combination of bounds. If, for instance, one set of information tells us that A is in a particular interval and another set tells us that A is in a different interval, it is straightforward to compute what the aggregate data set is implying simply by taking the intersection of the 2 intervals. When we have 2 estimates from separate approximations, on the other hand, we would have to invoke a much more complicated meta-analysis to combine the estimates.

In many cases, determining the correct decision does not require perfect precision. Analysis can reveal whether the uncertainty makes it unclear what the best decision is. Because the bounds on uncertainty tend to tighten as we collect more data, as soon as the best decision is obvious, one can stop gathering data. For approximations that contain no associated statement about their own reliability, scientists tend to always clamor for more data no matter how much they already have. For approximations, it takes an explicit uncertainty analysis to discern whether the data are essential to make the decision at hand.

6.1.2 Why Worst Case and Intervals Are Not Enough

The crudest form of bounding analysis is just interval arithmetic (Moore 1966; Neumaier 1990). In this approach the uncertainty about each quantity is reduced to a pair of numbers, an upper bound and a lower bound, that circumscribe all the possible values of the quantity. In the analysis, these numbers are combined in such a way to obtain sure bounds on the resulting value. Formally, this is equivalent to a worst case analysis (which tries to do the same thing with only 1 extreme value per quantity). The limitations of such analyses are well known. Both interval arithmetic and any simple worst case analysis

- Cannot take account of distributions
- Cannot take account of correlations and dependencies
- Can be hyperconservative
- Do not express likelihood of extremes

With an assessment as crude as interval or worst case analysis, it is impossible to make use of detailed empirical information, which is sometimes available, about a quantity beside its potential range. It wouldn't help, for instance, to know that most values are close to some central tendency, or that the variation in the quantity expressed through time follows a normal distribution. Knowledge about the statistical associations between variables is also useless in crude bounding analyses. For instance, if we know that large organisms drink more contaminated water than small ones and thus receive larger exposures, we may not be able to account for this association to gain more specific information about the final risk to the population. Because these methods do not use all the available information, they can produce results that are more conservative than is necessary given what is known. This leads to wasted resources in cleanup or unnecessary restrictions on pesticide applications. The central problem with crude bounding is that it addresses only the bounds on risks. It makes no statement about how likely such extreme risks are. Even if the upper bound represents an intolerable risk, if the chance of it actually occurring is vanishingly small, it may be unreasonable to base regulation on this value. Effective management requires that we somehow estimate these chances.

6.1.3   LIMITATIONS OF TRADITIONAL APPLICATIONS OF PROBABILITY

Probability theory is, of course, designed precisely to estimate these chances. Because of this, probabilistic assessment is regarded by many as the heir apparent to worst case analysis. However, traditional applications of probability theory also have some severe limitations. As it is used in risk assessments today, probability theory

* Requires a lot of information, or else subjective judgment
* Cannot address shape or model uncertainty
* Makes back calculation cumbersome or impossible
* Has an inadequate model of ignorance
* Confounds variability with incertitude

That probabilistic assessment requires a lot of data to parameterize is obvious to anyone who has attempted to use it. Once one has established the mathematical expression to evaluate, one must come up with estimates for the statistical distribution for each parameter in the expression. This implies knowledge not only of the means and variances, but in principle, of all of the moments and of all details about each distribution. One then is responsible for specifying the multivariate dependencies among all the parameters as well. This task is well beyond specifying all possible pairwise correlations (which, by itself, is so hard that virtually no one has the requisite empirical information). As a consequence of this need, analysts routinely make assumptions or use subjective judgments about quantitative details, which are, of course, difficult to justify when the assessment is scrutinized in public review.

There are also some technical problems with current probabilistic assessments. Notably, there is no satisfactory way to handle uncertainty about the proper mathematical model to use. (Is this the correct expression to compute in the first place? Are these assumptions appropriate?) Just as important is the lack of effective strategies for inverting the equations involving uncertainty to find solutions to various back calculation problems. (What distribution of environmental concentrations can be allowed given constraints on the resulting doses that can be tolerated in organisms?) Traditional probabilistic assessments conducted for pesticides have no robust and comprehensive strategies to answer these kinds of questions.

The fundamental limitation of probability theory, at least in the way it is usually applied in practical assessments, is that it has an inadequate model of ignorance. For instance, when all that is known about a quantity is its theoretical range, probabilists traditionally employ a uniform distribution over this range to represent this uncertainty. This approach dates back to Laplace himself and his "principle of insufficient reason." The approach is also justified by modern reasoning appealing to the "maximum entropy criterion" (Jaynes 1957; Lee and Wright 1994). But not knowing the value of a quantity is not the same as having it vary randomly. When probabilists do not distinguish between equiprobability and ignorance, they are confounding variability with incertitude.

6.1.4 DISADVANTAGES OF 2ND-ORDER MONTE CARLO SIMULATION

Some analysts suggest the use of 2nd-order probabilistic methods to overcome the limitations outlined above. The idea is to strictly separate variability from incertitude. Second-order Monte Carlo simulation is often offered as a way to effect this separation. Unfortunately, this approach is not without its own problems. Second-order Monte Carlo simulation

- Is expensive to compute (involving squared effort)
- Can be daunting to parameterize
- Has ugly displays that can be hard to explain
- Encounters some technical problems in computations
- Cannot handle shape or model uncertainty
- Is cumbersome in a back calculation
- Does not handle incertitude correctly

Second-order Monte Carlo simulation consists of a Monte Carlo simulation in which each iteration represents an entire Monte Carlo simulation. The calculation thus usually demands a squared calculational effort. Because 2nd-order Monte Carlo simulation requires the analyst to specify a distribution for each parameter of the primary distributions (and, one supposes, the ancillary dependencies as well), analysts who were already heavily taxed in simple simulations can find it overwhelming to parameterize the full 2-dimensional effort. The graphical displays that result from 2nd-order Monte Carlo simulations are sometimes called by the evocative names of "spaghetti plots" or even "bad hair day plots." These outputs can be confusing to interpret even for professionals and often engender distrust or even laughter among nontechnical decision makers to whom they are shown.

There are also some technical difficulties with this approach. For instance, if the distributions for the minimum and maximum of a uniform distribution overlap at all, then there is a possibility that the selected minimum exceeds the selected maximum. It is unclear what analysts should do in such a situation. The problem is actually rather general, because, for many statistical distributions, there are logical constraints that govern the relations between the parameters (see Frey and Rhodes 1998). Other technical problems include the fact that this approach, like simple Monte Carlo simulation, has no comprehensive strategy for addressing model uncertainty or solving back calculations. Although back calculations have been attempted in the context of 2nd-order simulations (cf. Parysow and Tazik 2001), they are at best rather cumbersome because they must be done via trial and error with trials that are computationally expensive to start with.

Finally, we argue that this approach does not handle incertitude correctly. Although a 2nd-order Monte Carlo simulation does conscientiously separate variability from incertitude, it still applies the methods of Laplace to incertitude. As a result, it can produce estimates that are inappropriate or unusable in risk analysis (Ferson and Ginzburg 1996).

6.1.5   WHAT IS NEEDED?

We suggest that what is needed is a bounding approach that marries the advantages of interval analysis with those of probability theory while sidestepping the limitations of both. In the following sections, we describe 2 very different approaches that do just this in different ways.

6.2   ROBUST BAYES

In a regular application of Bayes's rule, a prior estimate of probability and a likelihood function are combined to produce a posterior estimate of probability, which may then be used as an input in a risk analysis. Bayes's rule is

$$p(\theta|E) = p(\theta)\, p(E|\theta)/p(E)$$

where p denotes probability mass or density, θ is the value of the quantity in question, E denotes the evidence being considered, $p(\theta)$ is the prior probability for a value θ, $p(E|\theta)$ is the conditional likelihood function that expresses the probability of the evidence given a particular value of θ, and $p(E)$ is the probability of having obtained the observed evidence. For most Bayesians, the prior estimate represents the opinion or belief of the analyst, obtained through reflection or self-examination. It is intended to represent, at least initially, the analyst's subjective knowledge before any specific evidence is considered. It may be the result of amorphous preconceptions or mechanistic reasoning or a combination of the 2. It may also be the posterior from an earlier application of Bayes's rule conducted with a separate collection of data from a previous prior estimate. The likelihood function represents a model, also perhaps taken from the subjective knowledge of the analyst, of what data would imply about the variable in question. Traditionally, Bayesians assume ideal precision, that is, that both the prior and the likelihood are perfectly well-specified probabilities.

When there are many possible values of θ and the prior $p(\theta)$ is a probability distribution and the likelihood function $p(E|\theta)$ is defined on the same axis,

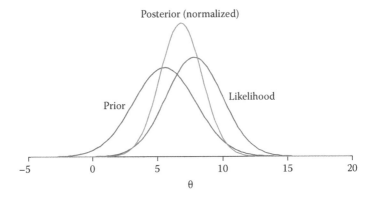

FIGURE 6.1   Bayes combination of a prior distribution and a likelihood function to obtain a posterior distribution for θ. The vertical axis (not shown) is probability density.

then Bayes's rule can be used to obtain a probability distribution for the posterior $p(\theta|E)$. This is done by applying the rule for each value of θ. For example, Figure 6.1 depicts the application of Bayes's rule to a prior distribution to obtain a posterior distribution. The probability of the evidence $p(E)$ in such cases is the same for all values of θ and is sometimes called the normalization factor because it's the divisor that makes the posterior distribution end up with unit area. This divisor is the sum or integral with respect to θ of the product of the prior and the probability of observing a value if the value were actually θ. The rule is applied for all values of θ to obtain $p(\theta|E)$, which is the distribution of θ given the evidence. When the normalization factor is computed via an integral expression, the computational burdens associated with applying Bayes's rule for distributions can be substantial. There is usually no closed-form solution available for computing the integral in the denominator, unless the prior and likelihood happen to constitute a "conjugate pair" for which the analytical details work out nicely. For instance, under particular assumptions, the following pairs of prior and likelihood (from which observations are drawn) yield the posterior distribution shown in Table 6.1.

TABLE 6.1
Posterior distribution

Prior	Likelihood	Posterior
Beta	Bernoulli	Beta
Beta	Binomial	Beta
Gamma	Poisson	Gamma
Beta	Negative binomial	Beta
Normal	Normal	Normal
Gamma	Normal	Gamma
Inverse-gamma	Exponential	Inverse-gamma

For these pairs, updating rules permit the immediate specification of the posterior's parameters from those of the prior and statistics from the data. For the assumptions underlying the use of these conjugate pairs and details on exactly how the calculations are to be made, consult standard references on Bayesian methods (e.g., DeGroot 1970; Berger 1985; Sander and Badoux 1991; Gelman et al. 1995; Lee 1997). Naturally, the existence of these conjugate pairs greatly simplifies the demands of applying the rule and are widely used for the sake of convenience, but of course they are very restricted in scope and obviously require distributional assumptions.

Robust Bayes methods (Berger 1985; Insua and Ruggeri 2000) acknowledge that it is sometimes very difficult to come up with precise distributions to be used as priors. Likewise, the appropriate likelihood function that should be used for a particular problem may be in doubt. In a robust Bayesian analysis, a standard Bayesian analysis is applied to a prior distribution and a likelihood function selected from *classes* of priors and likelihoods considered empirically plausible by the analyst. This approach has also been called "Bayesian sensitivity analysis." It is depicted in Figure 6.2, in which a class of priors and a class of likelihoods together imply a class of posteriors by pairwise combination through Bayes's rule. A result is said to be robust if it's approximately the same for each such pair. If the answers differ substantially, then their range is taken as an expression of how much (or how little) can be confidently inferred from the analysis. Robust Bayes also uses a similar strategy to combine a class of probability models with a class of utility functions to infer a class of decisions. A decision analysis is robust if all the possible combinations lead to the same decision. Because robustness reflects the insensitivity of the quantitative result to small changes in the underlying assumptions of the analysis, it has important relevance for the overall reliability of the analysis whenever those assumptions are tenuous or in contention.

The classes specified in a robust Bayesian analysis can be defined in a variety of ways, depending on the nature of the analyst's uncertainty. For instance, one could specify parametric classes of distributions in one of the conjugate families (e.g., all the beta distributions having parameters in certain ranges). Alternatively, one could specify parametric classes of distributions but not take advantage of the conjugacies.

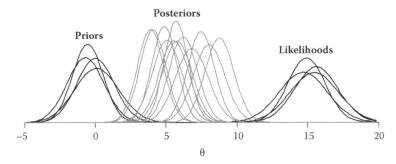

FIGURE 6.2 Robust Bayes combination of several prior distributions and likelihood functions to obtain many possible posterior distributions. The vertical axis (not shown) is probability density.

The subsequent calculations would probably be a lot harder in this case. There are also various other approaches based on density ratios (bounded density distributions), ε-contamination models, mixtures, quantile classes, and bounds on cumulative distribution functions. See Berger (1985, 1994) for an introduction to these ideas.

6.2.1 NUMERICAL EXAMPLE

Suppose that a prior distribution is within the class of all normal distributions having a mean between −1 and +1 and a variance between 1 and 2.5. Suppose further that the likelihood function is also characterized by a normal shape, with an unknown mean observed to be in the interval [14, 16] and specified variance in the interval [1.7, 3]. In Figure 6.2, several prior distributions and likelihood functions from their respective classes are drawn on a θ-axis in terms of probability density. Also shown on the same axis are several representatives of the class of posterior distributions that are obtained by applying Bayes's rule to every possible pair of prior distribution and likelihood function. These are shown as gray curves. Because the priors and likelihoods are conjugate pairs, it is easy in this example to compute the posteriors, which will also be normally distributed. The result is a class of normal distributions having a mean in the interval [2.75, 9.93] and a variance in the interval [0.63, 1.36]. This class of posteriors reflects the incertitude of the result given the professed incertitude about the prior and the likelihood. (The wide discrepancy in this example between the priors and the likelihoods was used so the reader could visually distinguish the 3 classes of curves. One might typically expect the priors and the likelihoods to overlap much more broadly.) The details of this numerical example depend in part on the use of normality assumptions and could differ if other assumptions were made.

6.2.2 ADVANTAGES OF THE APPROACH

In robust Bayesian analysis, the insistence on having a single, precise prior distribution and a single, specific likelihood function is relaxed. In their places, entire classes of distributions and functions are used. Although this approach is clearly inconsistent with the Bayesian idea that uncertainty should be measured by a single additive probability measure and that personal attitudes and values should always be measured by a precise utility function, the robust approach can be justified as a matter of convenience because arriving at precise statements that encapsulate an analyst's beliefs can be difficult and time consuming. Some analysts also suggest that robust methods extend the traditional approach by recognizing incertitude as a different kind of uncertainty (Insua and Ruggeri 2000; cf. Berger 1994).

Robust Bayes redresses some of the most commonly heard criticisms of the Bayesian approach. For instance, robust Bayes relaxes the requirement for an analyst to specify a particular prior distribution and reflects the analyst's confidence about the choice of the prior. Bayesian methods generally preserve zero probabilities. That is, any values of the real line for which the prior distribution is surely zero will remain with zero probability in the posterior, no matter what the likelihood is and no matter what new data may arrive. This preservation of zero probabilities means that an erroneous prior conception about what is possible is immutable in the face of any new evidence or argument. In the case of robust Bayesian analysis, only those regions of the real line where all the prior distributions in the permissible class are identically zero would be preserved at zero.

6.2.3   LIMITATIONS OF THE APPROACH

There are, however, other limitations of Bayesian methods that are not obviated or relaxed by a robust approach. The zero-preservation problem is an extreme case of a more general problem of Bayesian methods having to do with their possible insensitivity to surprise in the form of unanticipated data (Hammitt 1995). For instance, in the numerical example above, the posterior substantially disagrees with both the prior and the new data. When expectation and evidence are in conflict, an analyst might prefer fidelity to either one or the other, rather than seeming to split the difference to a compromise neither supports. Another somewhat troublesome feature of Bayesian calculations illustrated by the example is that, despite the apparently surprising nature of the evidence, the posterior distributions can be tighter (that is, have smaller variance) than the prior distributions. In the case of extreme surprise such as this, one might prefer a result that represented more uncertainty, rather than less.

Another potentially serious limitation of robust Bayes methods is that their computational costs can be large. The complexity of the requisite calculations depends on how the class of priors and the class of likelihoods are specified. In some cases, the use of computers may lessen the burden on human analysts, although there does not yet exist convenient software for this purpose.

The definition of the classes that characterize one's uncertainty about the correct prior distribution or likelihood function to use can be a rather subtle business. There are various ways to construct the classes. Berger (1994) suggests that desirable properties of the classes would be that they are

- Easy to understand and elicit
- Easy to compute with
- Sufficiently big to reflect one's uncertainty
- Generalizable conveniently to multiple dimensions

It is possible that one could specify a class that seems broad and yet does not really circumscribe the true uncertainty. For instance, suppose we define a class of prior distributions by reference to the parameters of a named probability distribution. This would be a very natural thing to do, for instance, if we wanted to use the conjugate pairs to simplify the calculations. We might talk about a class of normal distributions that all have the same variance but different means. A caricature of this class is shown in the left set of cumulative distribution functions depicted in Figure 6.3. Alternatively, we could talk about a class of prior distributions that all have the same mean but different variances (depicted on the right of the figure). We could also talk about the much larger class of all normal distributions having a mean in some range and a variance in some range. This class would be much harder to depict because it has so many elements. However, all of the elements are still very special because

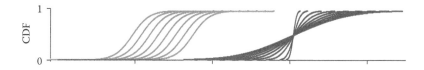

FIGURE 6.3 Two parametric classes of prior distributions having constant variance (left) or constant mean (right) shown as cumulative distribution functions (cdfs). The horizontal axis is some value for a random variable and the vertical axis is (cumulative) probability.

of their normality. The class, although it could be very wide if the ranges of the parameters are wide, is still extremely sparse in the sense that it excludes almost all distributions that have roughly similar distribution shapes. Being normal means that each distribution is perfectly symmetrical and balances its mass in the tails in a very special way. An analyst needs to decide whether the class is sufficiently rich to express the true breadth of uncertainty.

In some situations where uncertainty is great, an analyst might want to define classes that are nearly vacuous so that they say as little as possible about what the true prior and true likelihood function are. There is one important caveat, however, about specifying the classes too broadly. If the class of priors is specified only by bounds on the cumulative distribution function (cdf) and the class of likelihood functions is likewise specified only as bounds on cumulative probability, then all one can conclude about the posterior is its range (which turns out to be the intersection of the ranges of the prior and the likelihood). Thus, in this highly uncertain situation, a robust Bayesian analysis will always produce a trivial result that says essentially nothing about the class of posteriors. There is something of an art to picking the right classes that fully capture uncertainty but yet do not swerve into triviality.

6.3   PROBABILITY BOUNDS ANALYSIS

Probability bounds analysis is a related strategy for making probabilistic inferences in the face of incertitude. It is a method for computing bounds on the distribution of a sum, product, or arbitrary mathematical expression, given only bounds on the distributions of the addends, factors, or inputs. The bounds are expressed on cumulative distributions (rather than densities). This approach permits analysts to make risk calculations without requiring overly precise assumptions about parameter values, dependence among variables, or distribution shape. Probability bounds analysis gives the same answer as interval analysis does when only range information is available. It also gives the same answers as Monte Carlo analysis does when information is abundant enough to precisely specify input distributions and their dependencies. Thus, it is a generalization of both interval analysis and probability theory. In summary, probability bounds analysis

- Distinguishes variability and incertitude
- Makes use of available information
- Supports all standard mathematical operations

- Is computationally faster than Monte Carlo
- Is guaranteed to bound answer
- Often produces optimal solutions

6.3.1 What Is a P-Box?

Probability bounds analysis takes as inputs structures called "p-boxes," which express sure bounds on a cumulative distribution function. One p-box is depicted in Figure 6.4.

The upper and lower bounds touch at the point 0.25 for $x = 20$. The bounds are also coincident for values of x below 10 and above 60. P-boxes need not be composed of step functions such as shown in Figure 6.4. The bounds may be smooth or, indeed, can be any shape as long as they are monotonically increasing and do not cross each other. A p-box is designed to simultaneously express both variability and incertitude. Probability distributions, intervals, and scalar numbers are all special cases of p-boxes. Because a probability distribution expresses variability and lacks incertitude, the upper and lower bounds of its p-box are coincident for all x values at the value of the cumulative distribution function (which is a nondecreasing function from 0 to 1). An interval expresses only incertitude. Its p-box looks like a rectangular box whose upper and lower bounds jump from 0 to 1 at the endpoints of the interval. A precise scalar number lacks both kinds of uncertainty. Its p-box is just a step from 0 to 1 at the value of x corresponding to the scalar value.

There is a duality in the way that a p-box can be interpreted. It can be understood as bounds on the cumulative probability associated with any x value. For instance, in the p-box depicted in Figure 6.4, the likelihood the value will be 15 or less is between 0 and 25%. A p-box can also be understood as bounds on the value at any particular probability level. In the figure, the 95th percentile is sure to be between 40 and 60.

In risk analyses, p-boxes serve as models of the total uncertainty about individual variables. There are several ways to obtain p-boxes from data and analytical judgment. But, before we consider where p-boxes come from, let's first review what we can do with them, in particular, how we can use p-boxes in risk calculations.

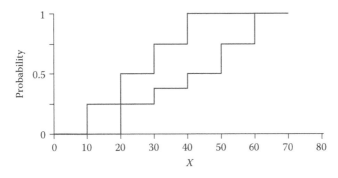

FIGURE 6.4   A probability box or p-box.

6.3.2 How Do You Compute with P-Boxes?

Probability bounds analysis combines p-boxes together in mathematical operations such as addition, subtraction, multiplication, and division. This is an alternative to what is usually done with Monte Carlo simulations, which usually evaluate a risk expression in one fell swoop in each iteration. In probability bounds analysis, a complex calculation is decomposed into its constituent arithmetic operations, which are computed separately to build up the final answer. The actual calculations needed to effect these operations with p-boxes are straightforward and elementary. This is not to say, however, that they are the kinds of calculations one would want to do by hand. In aggregate, they will often be cumbersome and should generally be done on computer. But it may be helpful to the reader to step through a numerical example just to see the nature of the calculation.

Suppose we have 2 p-boxes corresponding to 2 random variables, say A and B, shown in Figure 6.5, and we wish to compute bounds on the distribution of the sum $A + B$.

The 1st step is to partition the p-boxes of the addends into sets of intervals and associated probability masses. The p-box for A can be partitioned into the following 3 interval-mass pairs:

$$A \in [1, 2], \text{prob} = 1/3$$

$$A \in [2, 4], \text{prob} = 1/3$$

$$A \in [3, 5], \text{prob} = 1/3$$

(The symbol $\in$ is read "is an element of." When we write $A \in [1, 2]$, we mean that the value of A is some value between 1 and 2, inclusive.) This partitioning means that the probability is one-third each a value of the random variable is in each of the 3 intervals. In fact, this p-box could also be partitioned into more pairs. For instance, we could divide the 1st pair into 2, such as ([1, 2], 1/6) and ([1, 2], 1/6). This would make no essential difference in the calculation, nor have any consequence for the final result. Whatever the partition, the sum of the masses must be unity. A natural partition for the 2nd p-box in Figure 6.5 is

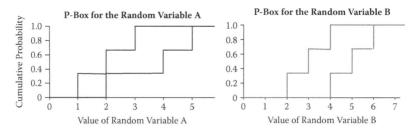

FIGURE 6.5 P-boxes for uncertain random variables A and B.

$$B \in [2, 4], \text{prob} = 1/3$$

$$B \in [3, 5], \text{prob} = 1/3$$

$$B \in [4, 6], \text{prob} = 1/3$$

We are now ready to combine the 2 p-boxes. To do so, we construct the Cartesian product of the 2 collections of interval-mass pairs in the matrix shown in Table 6.2.

For each cell inside this matrix, there is an interval, which is the range of possible values for the sum given the ranges of the marginal intervals for A and B, and a probability, which is the product (under independence) of the 2 marginal probabilities. Notice that the elements inside the matrix are another collection of intervals with associated probability masses. Because the probabilities add up to 1, they also specify a p-box. Figure 6.6 shows this p-box reassembled from the 9 interval-mass pairs in the matrix.

It turns out that, given the variability and incertitude in the inputs, this is the best possible p-box for the sum $A + B$. This means that we could not tighten the bounds in any way and still have it include all possible distributions that could arise as a sum of distributions from inside the p-boxes of the inputs.

TABLE 6.2
Cartesian product of 2 collections of interval–mass pairs

$A + B$ independence	$A \in [1, 2]$ prob = 1/3	$A \in [2, 4]$ prob = 1/3	$A \in [3, 5]$ prob = 1/3
$B \in [2, 4]$ prob = 1/3	$A + B \in [3, 6]$ prob = 1/9	$A + B \in [4, 8]$ prob = 1/9	$A + B \in [5, 9]$ prob = 1/9
$B \in [3, 5]$ prob = 1/3	$A + B \in [4, 7]$ prob = 1/9	$A + B \in [5, 9]$ prob = 1/9	$A + B \in [6, 10]$ prob = 1/9
$B \in [4, 6]$ prob = 1/3	$A + B \in [5, 8]$ prob = 1/9	$A + B \in [6, 10]$ prob = 1/9	$A + B \in [7, 11]$ prob = 1/9

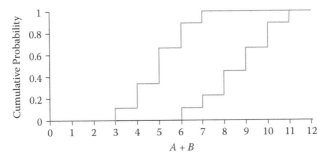

P-Box for $A + B$ Assuming Independence

FIGURE 6.6 Sum of the uncertain numbers depicted in Figure 6.5.

Obviously, we've picked the inputs for this example so that they would partition easily into small collections of paired intervals and probabilities. When the input p-boxes are smooth or complicated continuous structures, the partitions are discretizations that will often need to be much finer. The precision of the result depends on the fineness of these partitions, which can be arbitrarily increased to achieve whatever precision is required.

6.3.3   WHAT ABOUT CORRELATION AND OTHER DEPENDENCIES?

This calculation was performed under the assumption of mutual independence between the 2 variables. What about other sorts of dependencies? Excluding the case where one variable is a direct function of the other (which should be modeled directly in the assessment), the following statistical situations are commonly encountered in analyses:

- Independence
- Perfectly positive relation (maximal correlation)
- Opposite relation (minimal correlation)
- Particular nonlinear dependence (copula)
- Particular correlation coefficient
- Positively associated
- Negatively associated
- Unknown dependence

Convolutions with the first 4 of these dependencies can be computed with a table calculation similar to that illustrated in the previous section (Williamson and Downs 1990; Berleant 1993, 1996; Ferson and Long 1995). These dependencies introduce no further incertitude beyond what is already expressed by the input p-boxes. This means that if the inputs are precise probability distributions, the output will be too. In the case of the last 4 dependencies, on the other hand, further incertitude is introduced by not specifying the dependence fully. Thus, even if we start with precise distributions, the result will be a p-box. This is true even when we know a precise correlation coefficient. The reason is there are actually many different dependencies that correspond to any particular correlation coefficient. The last case of not knowing anything about the dependence is surprisingly common in risk assessments. Indeed, it is rare that available information includes the paired data necessary to make empirical statements about dependencies. Fortunately, Frank et al. (1987) showed how to compute best possible bounds on the distribution of sums and similar operations given only information about the marginal distributions without information about their dependence. Williamson and Downs (1990) showed how to extend these calculations to p-boxes to obtain sure bounds without making any assumption about the dependence between the variables. Their algorithms were not based on a table approach, but Berleant and Goodman-Strauss (1998) described such a table algorithm that used mathematical programming to obtain the same best possible bounds.

6.4 NUMERICAL EXAMPLE

In this section, we describe an exposure assessment for a hypothetical contact avicide involving 4 exposure pathways: maternal transfer A, ingestion B, imbibition (drinking) C, and dermal absorption D. In this example, we compute bounds on the sum $A + B + C + D$ from only partial information about each of the respective random variables. This example illustrates that it is easy to mix very different kinds of knowledge together in a bounding analysis. In this example, suppose that the distributional shape of A is known, but its parameters are in doubt. Suppose that a few parameters of B are known, but the shape or family of the statistical distribution is not known. Further suppose that sparse data were used to form the 95% confidence limits for the distribution of C. And the variable D is known to be well described by a precise distribution. Shown in Figure 6.7 are the bounds on each of the 4 inputs and bounds on the sum, both with an assumption of independence and without any assumption about the dependence among the variables. The dotted curves represent the inputs and answers that might have been associated with a traditional probabilistic assessment that did not acknowledge the uncertainty about the distributions and dependencies. Compare them with the solid edges of the p-boxes to quantify how much the tail risks would have been underestimated.

Table 6.3 lists the summary statistical measures yielded by 3 analyses of this hypothetical calculation. The 2nd column gives the results that might be obtained by a standard Monte Carlo analysis under an independence assumption (the dotted lines in Figure 6.7). The 3rd and 4th columns give results from probability bounding analyses, either with or without an assumption of independence.

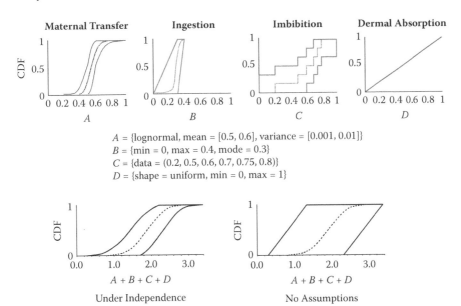

A = {lognormal, mean = [0.5, 0.6], variance = [0.001, 0.01]}
B = {min = 0, max = 0.4, mode = 0.3}
C = {data = (0.2, 0.5, 0.6, 0.7, 0.75, 0.8)}
D = {shape = uniform, min = 0, max = 1}

FIGURE 6.7 Example calculation of a sum of 4 addends characterized by p-boxes.

TABLE 6.3
Summary statistical measures resulting from hypothetical calculations

Summary	Monte Carlo	Independence	General
95th percentile	2.45	[2.1, 2.9]	[1.3, 3.2]
Median	1.87	[1.4, 2.3]	[0.79, 2.8]
Mean	1.88	[1.4, 2.3]	[1.4, 2.3]
Variance	0.135	[0.086, 0.31]	[0, 0.90]

Notice that, while the Monte Carlo simulation produces point estimates, the bounding analyses yield intervals for the various measures. The intervals represent sure bounds on the respective statistics. They reveal just how unsure the answers given by the Monte Carlo simulation actually were. If we look in the last column with no assumption, for instance, we see that the variance might actually be over 6 times larger than the Monte Carlo simulation estimates.

6.5   HOW TO USE BOUNDING RESULTS

Risk assessment is often marshaled in support of decision making. The kinds of decisions commonly addressed include questions about whether to regulate a new chemical or ban an old one, or how much to clean up a contaminated site. How will the results of bounding analyses be used in decision making? When the uncertainty makes no difference to the decision (because the results are clearly high or low), the bounding analysis gives confidence in the reliability of the decision. When, on the other hand, uncertainty makes the result so wide that the proper decision to make is obscured, one has 2 options. The first option is to use the results to demonstrate to managers which inputs need to be studied further to reduce the uncertainty enough to make a decision of appropriate reliability. When, as is the case in the numerical example above, the results are known to be best possible in the sense that they could not be tightened without further empirical information or theoretical assumption, then the argument for collecting further data is bolstered to the strongest it could possibly be. The manager can be shown that which decision is best is not knowable without the needed empirical investment.

The 2nd option when uncertainty swamps the decision is to use a secondary criterion to make the decision based on the possibilities within the probabilistic bounds. For instance, in environmental regulation, one may want to be conservative and ask about the worst case scenario. In the numerical example above, because large values of the sum represent adverse conditions, then one may choose to look at the right tail of the p-box and plan for the worst it suggests. This would mean, in this case, that we would act as though the right bounds of intervals estimating the mean and 95th percentile, etc., give their true values.

In other cases, one may elect to use the best case scenario and make plans based on the left bound within the p-box. Which criterion one might use is outside the scope of probability bounds analysis. However, it should be emphasized that the possibilities within the bounds are not equivalent. And the analyst should not pick any answer from within these bounds. We recall the case of the engineers who designed Kansai International Airport on an island constructed with fill in a harbor near Kobe, Japan. They were reportedly told by geologists that the island would settle between 19 and 25 feet. They chose to plan for 19-foot subsidence, supposedly because planning for 25 feet would have been prohibitively expensive. One needs not be a student of Greek tragedy to anticipate the fate of such hubris.

6.6 SEVEN CHALLENGES IN RISK ANALYSES

Most of the rest of this chapter is devoted to reviewing how bounding methods can address the following challenges routinely faced by analysts in pesticide risk assessments:

- Input distributions unknown
- Measurement error large
- Censoring
- Sample sizes small
- Correlation and dependency ignored
- Mathematical structure questionable
- Back calculation very difficult

6.6.1 INPUT DISTRIBUTIONS UNKNOWN

As suggested in the numerical example above, p-boxes can be constructed for a variety of states of knowledge. The graphs in Figure 6.8 illustrate several cases. The ordinate for each graph is cumulative probability. The abscissa for each graph is the X value for the variable of interest. The depictions are analogous to and generalizations of cumulative distribution functions (cdfs).

In the top, left graph of Figure 6.8 we see a p-box for the case when we know, perhaps from mechanistic reasoning, that the variable should be log-normally distributed, but we can only give bounds on the mean and standard deviation for the distribution. In this case, we knew the mean had to be between 0.5 and 0.6, and the standard deviation had to be between 0.05 and 0.1. Obviously, when the intervals are very tight, the p-box becomes equivalent to a precise distribution. When the intervals are very wide, the p-box becomes broad. The p-box shown represents the best possible bounds. That is, the box is as tight as possible without excluding any log-normal distribution with the prescribed mean and standard deviation. Best possible p-boxes have been worked out for many distribution families. In particular, we can specify any of the following named distribution families and give intervals (or scalar numbers) to parameterize them:

Bernoulli	exponential	logistic	Rayleigh
beta	extreme value	log-normal	reciprocal
binomial	F (Fisher-Snedecor)	log-triangular	rectangular
Cauchy	Fréchet	log-uniform	Student's t
chi squared (X^2)	gamma	normal	trapezoidal
discrete uniform	Gaussian	Pareto	triangular
Dirac delta	geometric	Pascal	uniform
double exponential	Gumbel	Poisson	Wakeby
Erlang	Laplace	power function	Weibull

In general, this is the case for any distribution function for which one can compute the quantile (inverse distribution) function.

We can also construct a p-box if we can estimate the mean and some measure of dispersion, even if we have no idea at all what shape the distribution is or what distribution family it comes from. For instance, the p-box shown in the top, right graph depicted of Figure 6.8 illustrates the best possible p-box we'd get if we know the mean and standard deviation. The p-box is a consequence of the classical Chebyshev inequality. The middle, left graph depicts the best possible p-box when we know only the mean and the possible range of the variable. When we know the median, this "waist" pinches to a precise point at that value. This is because all distribution functions must pass through this particular point. Other cases, illustrated by the graphs in the bottom row of Figure 6.8, represent the case where we know

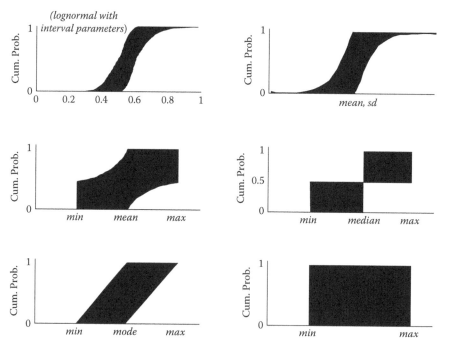

FIGURE 6.8 Several kinds of p-boxes for different states of knowledge about a random variable.

the variable's distribution is unimodal and the degenerate case, in the bottom right graph, of knowing only the minimum and maximum possible values of the variable. Contrast this box with the straight line that a uniform distribution would exhibit on the same axes.

In principle, one can fashion a p-box that represents the best possible limits on the distribution of a variable given any specific state of knowledge about the variable (Ferson 2002). Such optimal p-boxes have already been worked out for cases in which the following information is available. (Note that the values can be specified as precise scalar values or as interval bounds.)

- Minimum, maximum
- Minimum, maximum, mean
- Minimum, maximum, mode
- Minimum, maximum, any number of percentiles (e.g., median or 95th)
- Mean, 1 bound (minimum or maximum)
- Mean, dispersion (e.g., variance, standard deviation, coefficient of variation)
- Mean, dispersion, 1 bound (minimum or maximum)
- Minimum, maximum, mean, dispersion

The bounds can generally be tightened, sometimes considerably, by knowledge that the random variable is nonnegative or that the distribution function is symmetric or unimodal, or that it has a convex or concave hazard rate. The resulting p-boxes can also be combined using simple intersection as Rowe (1988) mentioned, although intersection does not necessarily guarantee that the result will continue to be the best possible result. It should be clear how using these kinds of p-boxes can allow an analyst to express uncertainty about input distributions — whether it is uncertainty about the parameters or uncertainty about the distribution shape — and propagate this uncertainty through the calculations of a risk assessment.

6.6.2 Measurement Errors Large

This and the next 2 subsections address how sample data can be used to construct p-boxes. Suppose that for a certain variable we have sampling data. These might be chemical concentrations measured in a laboratory. Suppose 15 such samples be represented as triangles distributed along an x-axis shown in Figure 6.9. The peaks of the triangles are the best estimates as point values, and their bases are the plus–minus ranges associated with the measurements. (In the illustration, we've shown the peaks to be centered over their bases, but this is not necessary.) The cumulative form of the empirical distribution function (edf) associated with these samples is shown as a gray stair-step function on the lower graph in Figure 6.9. It is formed by incrementing a step function by 1/15 at the location of each point value (triangle peak).

Also shown in black on the same scale is the p-box formed as 2 cumulative distribution functions, 1 based on the left endpoints of the triangle bases, and 1 based on the right endpoints. If the measurement errors associated with the samples are negligible, then the p-box will approach the gray edf. If measurement errors are large, the p-box will be wide. Measurement error, whether small or large, is almost

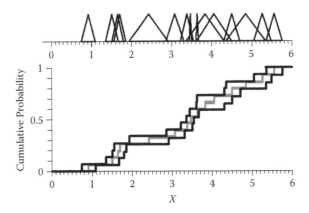

FIGURE 6.9 Empirical distribution function (gray, below) and p-box (black, below) corresponding to a data set (triangles, above) containing measurement error.

always ignored by analysts when they construct EDFs. Notice that the p-box, on the other hand, comprehensively expresses the measurement error exhibited in the sample data.

6.6.3 CENSORING

This construction of p-boxes is general enough to incorporate the uncertainty arising from data censoring. Suppose the laboratory that produced the data sample tells us that 4 of the 15 measurements were below detection limit. This means that, because of the dilutions they used and the analytical resolution of the devices they employed, they cannot be sure that the true values were not zeros.

The strategies used in traditional statistical analyses to handling such censoring range from simple substitution methods (e.g., replace each censored value by half the detection limit) to rather elaborate distributional models that attempt to reconstruct the now dubious values based on the patterns shown by the remaining values. Helsel (1990) reviews these strategies and points out the limitations of each. He notes that the current statistical methods

- Break down when censoring is prevalent
- Become cumbersome or unworkable with multiple detection limits
- Need assumptions about distribution shapes
- Yield approximations only

P-boxes, on the other hand, can readily express the uncertainty that arises from censoring, and they have none of these limitations mentioned by Helsel (1990). Suppose that 4 of the data values depicted in Figure 6.9 were identified by the lab as being below detection limit. Figure 6.10 shows how these data values would then be represented by triangles whose left endpoints are set to zero. The right endpoints are the respective detection limits. In this case, the detection limits are all 2.0, but they could

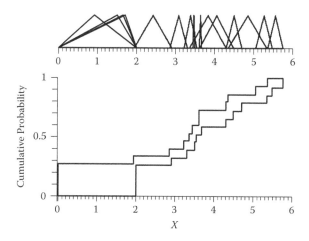

FIGURE 6.10 Empirical p-box (bottom) corresponding to a data set (triangles, top) with measurement error including 4 nondetect values.

be any value, and they could be different values for different estimates. The respective endpoints are then cumulated just as before. This p-box is trivial to compute, yet it obviously captures in a comprehensive way what censoring does. Moreover, this strategy can be used for right censoring as well as left censoring, or indeed for almost any kind of fundamental or happenstance limitation on mensuration.

In contrast to the limitations of traditional approaches to censoring, an approach based on p-boxes

- Works regardless of amount of censoring
- Handles multiple detection limits with no problem
- Makes no distribution assumptions
- Uses all available information
- Yields rigorous answers

Obviously, this kind of approach to data censoring does not result in a precise distribution, no matter how many data measurements are accumulated. By being conservative about measurement uncertainty, analysts can discern its consequences. If the effect of censoring is small, then the p-box will not be much wider on account of it.

6.6.4 Sample Sizes Small

In the sampling example considered in the previous 2 sections, if there are only 15 elements in the population, then forming the empirical bounding cumulative histograms as described above is a complete description of the uncertainty in that small population. The more typical situation, however, is that these 15 data values are just a small sample from a much larger population. If we collected another sample of measurements, the picture of variation and incertitude would probably be different. How should we account for sampling error that arises from measuring only a portion of the population? It would seem reasonable to inflate the uncertainty about the empirical histograms in some way.

The sampling theory for probability bounds analysis needs more development, but 1 strategy suggests itself. Kolmogorov–Smirnov (KS) confidence limits (Crow et al. 1960, p 90f; cf. Sokal and Rohlf 1980, p 721) are distribution-free bounds on a (precise) empirical distribution function as a whole. These limits are computed as $edf(x) \pm D_{\alpha,n}$ where edf is the empirical distribution function for any value x of the random variable, and $D_{\alpha,n}$ is the 1-sample Kolmogorov–Smirnov critical statistic for confidence level $100(1-\alpha)\%$ and sample size n. The values for D are tabled by Crow et al. (1960, Table 11 on p 248) and by Rohlf and Sokal (1981, table 32 on p 203). This formula can be extended to the p-boxes described in the previous sections that were formed by integrating left or right endpoints of plus–minus measurement intervals. For instance, the 95% KS confidence limits applied to the original interval sampling data are shown in black on Figure 6.11. It was derived from the gray p-box by simply adding and subtracting $D_{0.05,15} = 0.338$ to the upper and lower bounds, respectively. With only 15 data points, we would expect fairly low confidence in the precise empirical distribution function, but as the number of samples becomes large, the confidence limits get closer together. Note, however, that even for very large samples, the bounds cannot get closer than the incertitude from measurement error prescribes.

The KS limits make no distributional assumptions, but they do require that the samples are independent and identically distributed. Additional distributional assumptions can be made that could tighten the KS limits. For instance, assuming the underlying distribution from which the samples came is normal yields a much tighter p-box. In practice, the assumption about independence of the individual samples may sometimes be hard to justify, such as when contamination hotspots are the focus of targeted sampling efforts. Techniques to account for nonrandom sampling are a topic of current research.

The KS limits are certainly a standard idea in probability theory and have been used in traditional risk analyses, for instance as a way to express the reliability of the results of a simulation. However, it has not heretofore been possible to use KS limits to characterize the statistical reliability of the inputs. There has been no way to propagate KS limits through calculations. Probability bounds analysis allows us to do this

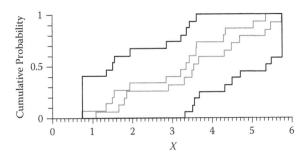

FIGURE 6.11   Kolmogorov–Smirnov confidence limits (black) accounting for both measurement uncertainty and sampling uncertainty about the p-box (gray) from Figure 6.9.

for the first time in a convenient way. Notice, however, that a p-box defined by KS confidence limits is different from the sure bounds formed by knowledge of moments or shape information that we discussed above. The KS bounds are not certain bounds, but statistical ones. The associated statistical statement is that 95% (or whatever) of the time the true distribution will lie inside the bounds. Using KS confidence limits as a p-box in probability bounds analysis amounts to assuming that the underlying unknown distribution is surely within the KS limits. Such an assumption is, in essence, no different than other assumptions analysts make in constructing p-boxes, such as that the moments or shape information is known or can be strictly bounded.

6.6.5 CORRELATIONS AND DEPENDENCIES IGNORED

As mentioned above, calculations with p-boxes can be made under an assumption of independence, assuming perfect or opposite dependence, or any specific dependence, or without making any assumption at all about the dependence. Figure 6.12 illustrates possible distributions of products AB, where A is uniformly distributed over the interval [2, 5] and B is normally distributed with mean 4 and standard deviation 1, under various assumptions about the dependence between A and B. If the variables are perfectly associated so that their correlation is maximal, i.e., as close as possible to +1 given the stated marginal distributions for A and B, then the cumulative distribution function for the product AB is the shallowest cumulative distribution in the figure (depicted with a dotted curve). If the correlation is opposite so it's as close to −1 as possible given the marginal distributions, then the distribution of the product is the steepest distribution (shown as a dashed curve). This cumulative distribution function has an interesting little hook around the value of 12. If the 2 variables are mutually independent, the distribution of their product is given by the cumulative distribution function shown with an intermediate slope (thin solid curve). The general bounds (thick solid curves), which make no assumption about the dependence between A and B, enclose all 3 distributions. As proven by Frank et al. (1987), these general bounds are the best possible bounds on

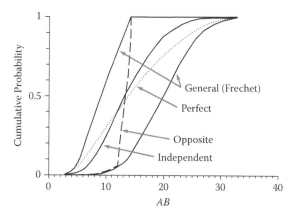

FIGURE 6.12 Different estimates about the distribution of products given different assumptions about the dependence between the factors.

the product. This means that these bounds could not be any tighter and still contain all the distributions that could arise under some dependency between *A* and *B*.

If an analyst conducted a sensitivity study to determine the effect of ignorance about the dependence between *A* and *B* by varying a correlation coefficient between +1 and −1, the results would sweep out the region between the perfect (dotted) and opposite (dashed) distributions. Notice however that this would not be sufficient to show the full range of uncertainty arising from such ignorance. That full range is given only by the enclosing (thick) general bounds and the methods of probability bounds analysis. In this case, the answer given by the p-box is both comprehensive and optimal.

6.6.6　Mathematical Structure of the Model Questionable

There are many situations in which even the correct form of the mathematical model to use in a risk analysis is in doubt. For instance, in population-level ecological assessments one must describe the prevailing density dependence that governs how population abundance approaches carrying capacity. There are several popular models, including the logistic model, Ricker function, Beverton–Holt function, ceiling model, and Shepard function. We do not have enough empirical information to distinguish among these possible models for most species of interest in ecological risk assessments. Thus there is model uncertainty that really ought to be propagated through an assessment. In practice, however, most analysts simply pick 1 of the density dependence models to use and ignore their uncertainty about the choice.

Some risk analysts have tried to address uncertainty about the model form within a Monte Carlo simulation (e.g., Morgan and Henrion 1990; Apostolakis 1995; cf. Cullen and Frey 1999). They first list a variety of possible models and then, inside the simulation, use a discrete random number to select which model from the list will be used in a particular simulation iteration. Within the iteration, this model is assumed to be true. This is repeated for many iterations, each time assuming a model selected at random from among those possible. Sometimes analysts weigh the selection of the discrete random variable according to their belief or judgment about likelihoods that the true model form is one or the other choices. This simulation strategy represents model uncertainty as a stochastic mixture of the possible models. In doing this, it effectively averages together incompatible theories. The approach is equivalent in this respect to the Laplacian approach to modeling what is fundamentally incertitude as an equiprobable stochastic mixture (the uniform distribution), and, as a result, it can underestimate the true tail risks in an assessment.

The probability bounding approach to this problem is to form the stochastic envelope of the possible models. For instance, suppose that we think that either model I or model II represents the fact of the matter, but we don't know which it is. Let's say these 2 models lead to 2 different distributions. Suppose they are the probability density functions labeled "I" and "II" in the upper graph of Figure 6.13.

If we cumulate each density function and form the envelope of their cdfs as shown in the lower graph in the figure, we create a p-box (labeled "I or II") that expresses the model uncertainty about whether model I or model II is the correct function to use. This approach is clearly more comprehensive than the traditional approach

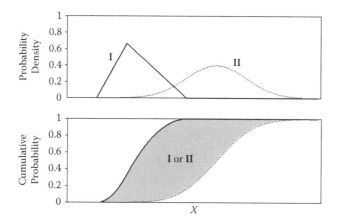

FIGURE 6.13 Use of a p-box (shaded region in lower graph) to represent uncertainty between models I and II summarized as distribution functions.

based on averaging of the 2 distribution functions (whose result is shown as the dark line within the shaded region of the lower graph). We note that the bounding approach can even handle nonstationarity of distributions, which is another important source of uncertainty that is usually ignored in traditional assessments for lack of a reasonable strategy to address it. This approach to model uncertainty works whenever the possible models imply different distribution functions for 1 or more variables. It works for more than 2 possible models, or even when there are infinitely many or innumerable possible models, or when the models cannot be explicitly listed. However, this approach can be overly conservative if the differences between the models are very great because it includes as possibilities a lot of intermediate models among the possible ones.

6.6.7   BACK CALCULATION VERY DIFFICULT

Planning remediation (cleanup) strategies often involves solving back calculation problems (Burmaster et al. 1995; Burmaster and Thompson 1995; Ferson 1995; Ferson and Long 1997). Back calculations are problems such as solving for B, given you know that $C = A \times B$, from an estimate of A and desired constraints on C. One common example is solving for limits on the distribution of environmental concentrations given some constraints on acceptable doses and the fact that dose = intake × concentration. Because the variables are not real numbers but uncertain quantities involving incertitude and/or variability, we cannot simply solve the equation using grade school algebra. To get the right answer, we need a special operation, called back calculation, that essentially untangles the convolution implied by the specified forward equation. Similar problems involving the untangling of operations other than multiplication or more complicated mathematical expressions composed of several operations are called back calculations too, and they also need special solution strategies.

The existing algorithms for doing back calculations or deconvolutions with probability distributions have notorious numerical problems (Jansson 1984). When given arbitrary inputs, such as might be defined by regulatory constraints, they usually crash and yield no answer at all. The problem is that probability distributions are overspecified. When A and C are precise distributions, it is usually the case that there simply is no distribution B that satisfies the equation. By design, p-boxes are a relaxation of the strictures of precise probability distributions. Because their interval nature relaxes the numerical problems, solutions to back calculation problems are easier to obtain for p-boxes. There is no guarantee that there will be a solution to an arbitrary equation involving uncertain numbers (i.e., quantities that harbor incertitude, variability, or both). But the severe algorithmic difficulties associated with precise distributions evaporate if the quantities are p-boxes, and this makes it far more likely that a solution exists.

P-boxes are also a much more natural way to express regulatory constraints in the first place. For instance, it is strange for regulators to offer a particular statistical distribution for exposures as the target for remediation or management (cf. Burmaster and Thompson 1995). Are they insisting that the remediation not be so effective that the frequencies of high exposures are lower than planned? Surely regulators would be even happier with any distribution whose graph is further to the left (representing decreases in all percentiles), just because it means that all exposures will generally be lower. Perhaps what a precise distribution really means when specified by regulators is an average target or an upper bound on the distribution of exposures. Viewed in this way, it is clear that p-boxes are the appropriate and natural way to express regulatory constraints.

6.7   WHAT BOUNDING CANNOT DO

There are 3 important limitations of probability bounds analysis. The 1st limitation is that, being only bounds on a distribution, a p-box cannot show what distribution is most likely within the box. A p-box provides no shades of gray or 2nd-order information that could tell us which distributions are the most probable. This is essentially the same problem, albeit at a higher level, that intervals had. It may, however, be possible to nest probability bounds analyses to get at the internal structure of the result. It is also often useful to simultaneously conduct a traditional Monte Carlo assessment, which will produce output distributions inside the output p-boxes. Together, these results characterize central and bounding estimates of the output distribution.

The 2nd limitation is that maintaining the optimality of answers may be hard when there are repeated variables or when there is a lot of empirical information about complex dependencies among the variables. Although the individual arithmetic operations yield best possible results at each step in the calculation, when these operations are chained together to compute the full risk expression, this optimality may be lost when variables appear multiple times in the expression, or when there are subtle intervariable dependencies present. Repeated variables are a problem essentially because they introduce their uncertainty more than once into the calculation. For this reason, the resulting bounds may not be as narrow as they should be. It is always possible to guarantee that the results will enclose the true result distribution, but to maintain the claim that the bounds are also best possible may require special calculation with brute-force algorithms.

The 3rd limitation is that all outputs must be expressed in terms of cumulative probability. It is usually not possible to depict results in terms of probability densities. This may not be a serious limitation, however, because, as reviewed in Morgan and Henrion (1990), pyschometric studies suggest humans are actually most facile at interpreting cumulative displays anyway. Nevertheless, some analysts may be annoyed that probability bounds cannot express results in terms of densities.

Robust Bayes methods share some of the same disadvantages of Bayesian methods in general. These include the tolerance of, or reliance on, subjective judgments. Although Bayesians regard this as an important feature of their approach, it has been hard to convince regulators that personal beliefs should play any role in assessments conducted to justify public policy. A persisting technical issue with Bayesian analyses is the zero-preservation problem, which is the total insensitivity of the posterior to data wherever the prior distribution is zero. In a sense, it is the mathematical analog of uneducable prejudice. By relaxing the focus from single distributions and precise functions, robust Bayes methods should tend to redress both of these disadvantages. The most serious limitation of robust Bayes methods is the practical one that no convenient software exists that makes it easy to apply in real-world problems. It is not always straightforward to specify or work with all the prior distributions (or likelihood or utility functions) in a class. This perhaps explains why there have not yet been any applications of robust Bayes methods to pesticide risk assessments.

Finally, although both probability bounds analysis and robust Bayes methods are fully legitimate applications of probability theory and, indeed, both find their foundations in classical results, they may be controversial in some quarters. Some argue that a single probability measure should be able to capture all of an individual's uncertainty. Walley (1991) has called this idea the "dogma of ideal precision." The attitude has never been common in risk analysis, where practitioners are governed by practical considerations. However, the bounding approaches may precipitate some contention because they contradict certain attitudes about the universal applicability of pure probability.

6.8   EXAMPLE: INSECTIVOROUS BIRDS' EXPOSURE TO PESTICIDE

Consider the following purely hypothetical example assessment for the exposure of an insectivorous bird to a new agricultural insecticide. This insecticide degrades very quickly after application, so that its toxicity dissipates after 24 hours. It does not bioaccumulate, and nonlethal doses are metabolized with no long-term consequences for the bird. These features imply a fairly simple assessment, involving the following expression for computing exposure of a bird to the insecticide within 1 day of its field application

$$\text{Dose} = \frac{\text{FIR} \times \text{Conc}}{\text{BW}} \text{Frac}$$

where FIR is the bird's food intake rate in grams per day, Conc is the concentration in the bird's diet in micrograms of insecticide per gram of insect tissue, BW is body mass of the bird, and Frac is the proportion of a bird's intake that occurs within the treated area.

There are abundant data on the body masses of the receptor birds, which are well modeled by a normal distribution with mean 14.5 g and a standard deviation of 3 g. No individuals smaller than 7 g or larger than 22 g have been observed. This distribution is depicted in the upper left graph in Figure 6.14. In this and all the graphs in this figure, the ordinate is exceedance risk (complementary cumulative probability). It is the chance that the random variable is as large as or larger than the value given on the abscissa. In the case of BW, this chance is known rather precisely for all possible values. It's convenient to show the p-boxes upside down like this when we want to focus on or emphasize the risks of high values.

Less is known about the distribution for food intake rate. One study reported in the literature suggests that the mean for daily consumption by this species in this region was 5.23 g insect tissue per day, with variance of 2.3 g. The published report suggested that these statistics were based on many observations, but it did not give any of the raw data or other information about the distribution. To model the

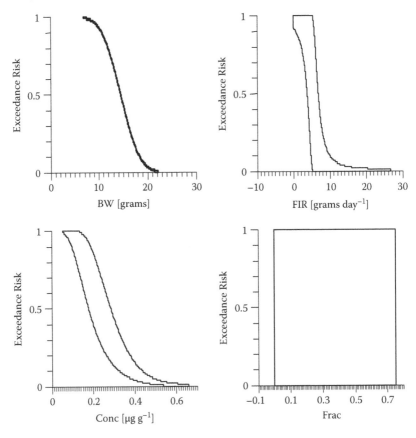

FIGURE 6.14 Input p-boxes for the example assessment of bird exposure to insecticide.

uncertainty about this variable, we can use the p-box shown in the upper right graph of Figure 6.14. It is the envelope of all possible (complementary) distribution functions for a positive random variable having the given mean and variance.

Fate and transport modeling was used to estimate the concentration of the insecticide in insect tissue consumed by birds. The details of this modeling effort, which we omit here, are rather complex and involve characteristics of the field application of the insecticide, local weather, multiple pathways of exposure to insects, sequestration of insecticide by mortality of insects, and integration over 0- to 20-g pools of insect tissue that would compose a bird's daily diet. The model of the pesticide's fate and transport made a prediction about the concentration variable, which is characterized by the p-box shown in the lower left graph of Figure 6.14. This p-box synthesizes all of the knowledge and uncertainty captured in the modeling effort. The model predicts the distribution function for concentrations, whatever it is, surely lies within the bounds shown.

The least well known variable in this assessment is Frac, the proportion of foraging a bird might do in a treated area. Local biologists admit that any estimate of it would be just a guess. Inferring from the foraging ranges observed for this species, they conclude that it could not be 100%, but that it could be as high as 75% for some or even most birds. They believe, however, that the lower bound for this variable could be zero. We represent this poor information as an interval depicted in the lower right graph of Figure 6.14. This interval is a degenerate p-box that represents all possible distribution functions over that range.

Analysts anticipate that these variables are statistically independent of each other, with the exception that FIR is likely to be positively correlated to BW, although no specific empirical evidence is available about the magnitude of this possible correlation. To account for this possibility, the quotient FIR/BW was first computed under the assumption that the 2 variables are positively correlated, but making no other assumption about their actual interdependency. The quotient was then multiplied by the p-boxes for Conc and Frac, assuming independence. The calculations were done with the Risk Calc software (Ferson 2002). The resulting p-box for the distribution of doses to birds is displayed in Figure 6.15. It suggests that doses are almost certainly smaller than 2 μg of insecticide per gram bird tissue over the course of the day following an application. The upper 95th percentile on such doses is surely less than 0.414 μg g^{-1} day^{-1}, and the median dose is in the interval [0, 0.142] μg g^{-1} day^{-1}. The mean dose is sure to be no larger than 0.092 μg g^{-1} day^{-1}. It could be as low as 0 (as would occur if Frac is 0). The standard deviation of doses is somewhere between 0 and 0.186 μg g^{-1} day^{-1}.

6.8.1 COMPARISON WITH A PRECISE ASSESSMENT

The insectivorous bird assessment can be compared to a more traditional probabilistic assessment based on precise distribution functions and particular dependence assumptions. For comparison purposes, we conducted such a simulation. The variable BW was modeled by the same normal distribution with mean 14.5 g and standard deviation 3 g. The variable FIR, on the other hand, was modeled by a log-normal distribution with mean 5.23 and variance 2.3 g per day. The choice of

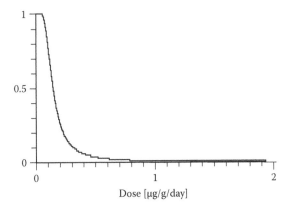

FIGURE 6.15   Output p-box for dose received by insectivorous birds.

the log-normal family reflected the analyst's best guess about the likely shape of the distribution but was not specifically supported by available empirical evidence. The distribution of Conc was also modeled with a log-normal distribution with mean 0.25 µg g^{-1} and standard deviation 0.1 µg g^{-1}. This choice obviously neglects the uncertainty about the distribution shape that was explicitly modeled for this variable. The parameter Frac was modeled with a uniform distribution ranging between 0 and 0.75. A uniform distribution is commonly used to represent interval uncertainty. All variables were assumed to be independent, except that BW and FIR were assumed to be perfectly correlated (Spearman rank correlation coefficient equal to 1). Of course these various modeling choices contain arbitrary elements, but these are necessary to specify the estimation problem completely.

The resulting distribution is shown as the black curve in Figure 6.16. For comparison, the p-box from Figure 6.15 is shown again in Figure 6.16 as a gray curve. The 95th percentile of the black dose distribution is 0.078 µg g^{-1} day^{-1}, and the median dose is 0.029 µg g^{-1} day^{-1}. Both of these values are about one-fifth of the upper estimates for the respective percentiles from the probability bounds analysis. In other words, given the professed uncertainty about the assessment, the true percentiles could be as much as 5 times larger than the values predicted by a Monte Carlo assessment based on unjustified assumptions. Likewise, the mean is 0.033 µg g^{-1} day^{-1} and the standard deviation is 0.025 µg g^{-1} day^{-1}. The upper limits on these statistics obtained by probability bounds analysis are roughly 2.5 and 7.5 times larger, respectively.

6.8.2   Back Calculation

The insectivorous bird assessment might also require a back calculation seeking a characterization of concentrations that could ensure the doses of insecticide received by the birds are no larger than can be physiologically tolerated. Suppose the toxicologists have collected evidence that doses lethal for the birds occur above 100 µg g^{-1} day^{-1}. Suppose also that regulators have concluded from this information that prudent environmental protection will require that all doses received by birds be less

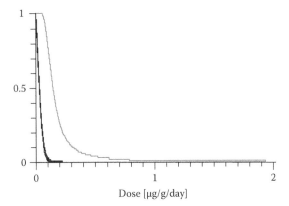

FIGURE 6.16   Output from traditional Monte Carlo assessment (black) compared to p-box (gray).

than 50 µg g^{-1} day^{-1}, and that at least 95% be less than 10 µg g^{-1} day^{-1}, and that the median dose received by birds be no more than 1 µg g^{-1} day^{-1}. (This conservativism is not an essential part of the example.) The p-box displayed on the left of Figure 6.17 circumscribes these constraints. Any complementary cumulative dose distribution lying entirely within this p-box will clearly satisfy the 3 constraints specified by the regulators. Given this p-box and the p-boxes in Figure 6.14 for the variables FIR, BW, and Frac, back calculation can compute a p-box that characterizes a set of distributions for Conc that will always be allowable. The details of this calculation are beyond the scope of this chapter, but the right p-box in Figure 6.17 displays its result. This p-box is interpreted as a "kernel" (rather than an envelope), which is to say that any distribution that lies entirely within the p-box will ensure that the resulting distribution of doses that arise from the environmental concentrations will surely satisfy the regulators' 3 constraints, no matter what the actual distributions for FIR and BW are within their respective p-boxes. There might be other concentration distributions that do not lie entirely within the kernel p-box that also satisfy

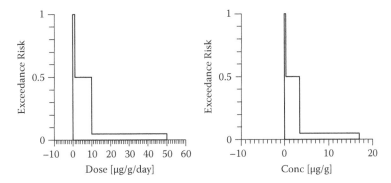

FIGURE 6.17   Planned constraints on the distribution of dose (left), and limits on the concentration distribution (right) that ensure dose constraints are not violated.

the constraints and yield a tolerable distribution of doses, but such distributions cannot be guaranteed a priori to do so given the expressed incertitude about the other variables. Tucker and Ferson (2003) describe back calculation with p-boxes and give algorithms to compute them.

6.9   CONCLUSION

Pesticide risk assessments must account for a variety of sources of uncertainty such as laboratory measurement error, small sample size, data censoring, and model uncertainty. The available techniques that account for these uncertainties in traditional probabilistic approaches are limited and usually require untestable assumptions. Bounding approaches have been developed to glean what can be reliably inferred from established scientific knowledge without recourse to unjustified assumptions, mathematical convenience, or wishful thinking. These bounding approaches are designed to retain the advantages of both probabilistic and worst case approaches to risk assessment but to sidestep their respective limitations.

6.10   APPENDIX

Following is the run stream used to make the calculations described in Section 6.8. Comments are delimited by double slashes. Output is shown in boldface type. Vertical bars around arithmetic operators mean they should be applied under an independence assumption.

```
// insectivorous birds' exposure to fast-decaying insecticide
// probability bounds analysis
BW = normal(14.5 grams, 3 grams)
FIR = posmeanstddev(5.23 grams per day, sqrt(2.3)*units("grams per day"))
Conc = lognormal([0.2,0.3] µg per g, 0.1 µg per g)
Frac = [0, 0.75]
Dose = convolution(FIR, BW, positive, divide) |*| Conc |*| Frac
Dose
```
   ~(range=[0,1.93614], mean=[0,0.092], var=[0,0.035]) day^{-1} µg g^{-1}
```
sd(Dose)
```
   [0, 0.1856] day^{-1} µg g^{-1}
```
cut(Dose, 95%)
```
   [0, 0.4145] day^{-1} µg g^{-1}
```
median(Dose)
```
   [0, 0.1423] day^{-1} µg g^{-1}
```
// compare against a precise probabilistic assessment
BW0 = normal(14.5 grams, 3 grams)
FIR0 = lognormal(5.23 grams per day, sqrt(2.3)*units("grams per day"))
Conc0 = lognormal(0.25 µg per g, 0.1 µg per g)
Frac0 = uniform(0, 0.75)
Dose0 = (convolve(FIR0, BW0, perfect, divide)) |*| Conc0 |*| Frac0
mean Dose0
```

  [**0.03325, 0.03336**] **day^{-1} µg g^{-1}**
  cut(Dose0, 95%)
  [**0.0739, 0.08288**] **day^{-1} µg g^{-1}**
  median(Dose0)
  [**0.02736, 0.02966**] **day^{-1} µg g^{-1}**
  // back calculation
  // lethal doses occur above 100 micrograms per gram per day
  MIN = 0 µg per g per day
  MAX = 50 µg per g per day
  MAX95 = 10 µg per g per day
  MAX50 = 1 µg per g per day
  Dose1 = constrainpercentiles(MIN, MAX, 50%, MAX50, 95%, MAX95)
  Conc1 = factor((FIR / BW) |*| Frac, Dose1)
  Conc1
  ~(range=[0,16.9585], mean=[0,2.55], var=[0,14.4]) µg g^{-1}

6.11   REFERENCES

Apostolakis GE. 1995. A commentary on model uncertainty. In: Mosleh A, Siu N, Smidts C, Lui C, editors. Proceedings of Workshop on Model Uncertainty. College Park (MD): Center for Reliability Engineering, University of Maryland.

Berger JO. 1985. Statistical decision theory and Bayesian analysis. New York: Springer-Verlag.

Berger JO. 1994. An overview of robust Bayesian analysis [with Discussion]. Test 3:5–124.

Berleant D. 1993. Automatically verified reasoning with both intervals and probability density functions. Interval Computations 2:48–70.

Berleant D. 1996. Automatically verified arithmetic on probability distributions and intervals. In: Kearfott B, Kreinovich V, editors. Applications of interval computations. Dordrecht (NL): Kluwer, p 227–244.

Berleant D, Goodman-Strauss C. 1998. Bounding the results of arithmetic operations on random variables of unknown dependency using intervals. Reliable Computing 4:147–165.

Burmaster DE, Lloyd KJ, Thompson KM. 1995. The need for new methods to backcalculate soil cleanup targets in interval and probabilistic cancer risk assessments. Human Ecol Risk Assess 1:89–100.

Burmaster DE, Thompson KM. 1995. Backcalculating cleanup targets in probabilistic risk assessments when the acceptability of cancer risk is defined under different risk management policies. Human Ecol Risk Assess 1:101–120.

Crow EL, Davis FA, Maxfield MW. 1960. Statistics manual with examples taken from ordnance development. New York: Dover Publications.

Cullen AC, Frey HC. 1999. Probabilistic techniques in exposure assessment. New York: Plenum Press.

DeGroot MH. 1970. Optimal statistical decisions. New York: McGraw-Hill.

Ferson S. 1995. Using approximate deconvolution to estimate cleanup targets in probabilistic risk analyses. In: Kostecki P. editor. Hydrocarbon contaminated soils. Amherst (MA): Amherst Scientific Press, p 239–248.

Ferson S. 2002. RAMAS risk calc software 4.0: risk assessment with uncertain numbers. Boca Raton (FL): Lewis Publishers.

Ferson S, Ginzburg LR. 1996. Different methods are needed to propagate ignorance and variability. Reliability Eng Syst Saf 54:133–144.

Ferson S, Long TF. 1995. Conservative uncertainty propagation in environmental risk assessments. In: Hughes JS, Biddinger GR, Mones E, editors. Environmental toxicology and risk assessment, 3rd vol, ASTM STP 1218. Philadelphia: American Society for Testing and Materials, p 97–110.

Ferson S, Long TF. 1997. Deconvolution can reduce uncertainty in risk analyses. In: Newman M, Strojan C, editors. Risk assessment: measurement and logic. Ann Arbor (MI): Ann Arbor Press.

Frank MJ, Nelsen RB, Schweizer B. 1987. Best-possible bounds for the distribution of a sum — a problem of Kolmogorov. Probability Theory and Related Fields 74:199–211.

Frey HC, Rhodes DS. 1998. Characterization and simulation uncertain frequency distributions: effects of distribution choice, variability, uncertainty, and parameter dependence. Human Ecol Risk Assess 4:423–468.

Gelman A, Carlin JB, Stern HS, Rubin DB. 1995. Bayesian data analysis. Boca Raton (FL): CRC Press.

Hammitt JK. 1995. Can more information increase uncertainty? Chance 36:15–17.

Helsel DR. 1990. Less than obvious: statistical treatment of data below the detection limit. Environ Sci Technol 24:1766–1774.

Insua DR, Ruggeri F, editors. 2000. Robust Bayesian analysis, lecture notes in statistics, vol 152. New York: Springer-Verlag.

Jansson PA, editor, 1984. Deconvolution with applications in spectroscopy. Orlando (FL): Academic Press.

Jaynes ET. 1957. Information theory and statistical mechanics. Phys Rev 106:620–630.

Lee PM. 1997. Bayesian statistics: an introduction. London: Edward Arnold (Hodder Arnold).

Lee RC, Wright WE. 1994. Development of human exposure-factor distributions using maximum-entropy inference. J Exposure Anal Environ Epidemiol 4:329–341.

Moore RE. 1966. Interval analysis. Englewood Cliffs (NJ): Prentice-Hall.

Morgan MG, Henrion M. 1990. Uncertainty: a guide to dealing with uncertainty in quantitative risk and policy analysis. Cambridge (UK): Cambridge University Press.

Neumaier A. 1990. Interval methods for systems of equations. Cambridge (UK): Cambridge University Press.

Parysow P, Tazik DJ. 2001. Assessing the effect of estimation error on population viability analysis: an example using the black-capped vireo. Engineer Research and Development Center, U.S. Army Corps of Engineers. Available from: http://handle.dtic.mil/100.2/ADA394828

Rohlf FJ, Sokal RR. 1981. Statistical tables. New York: Freeman.

Rowe NC. 1988. Absolute bounds on the mean and standard deviation of transformed data for constant-sign-derivative transformations. SIAM J Sci Stat Comput 9:1098–1113.

Sander P, Badoux R, editors. 1991. Bayesian methods in reliability. Dordrecht (NL): Kluwer.

Sokal RR, Rohlf FJ. 1980. Biometry. New York: Freeman.

Tucker WT, Ferson S. 2003. Setting cleanup targets in a probabilistic assessment. In: Mishra S, editor. Groundwater quality modeling and management under uncertainty. Reston (VA): American Society of Civil Engineers.

Walley P. 1991. Statistical reasoning with imprecise probabilities. London: Chapman and Hall.

Williamson RC, Downs T. 1990. Probabilistic arithmetic I: numerical methods for calculating convolutions and dependency bounds. Int J Approximate Reasoning 4:89–158.

7 Uncertainty Analysis Using Classical and Bayesian Hierarchical Models

D. R. J. Moore, W. J. Warren-Hicks,
S. Qian, A. Fairbrother, T. Aldenberg,
T. Barry, R. Luttik, and H.-T. Ratte

7.1 INTRODUCTION

Uncertainty is a term that embraces a variety of concepts (Morgan and Henrion 1990). It may arise because of vaguely stated policy goals, e.g., continued use of pesticides should not affect the sustainability of raptor populations. Uncertainty may arise because of differences in our preferences, e.g., what exactly is "acceptable" risk. It may refer to lack of knowledge about model structure, e.g., how should pesticide intake via preening be estimated, or lack of knowledge about a quantity like Henry's law constant for pesticide X. Uncertainty can sometimes be combined with the concept of variability, e.g., food intake rates among individuals in a flock of birds. Uncertainty also exists at different scales (e.g., spatially, temporally, levels of biological organization, etc.). These and other types of uncertainty can generate considerable confusion and often rancorous debate. As a result, there have been several attempts to classify types or sources of uncertainty (Finkel 1990; McNeill and Freiberger 1993; Hoffman and Hammonds 1994; Rowe 1994; Smith and Shugart 1994). Morgan and Henrion (1990) argue that it is crucial to distinguish between different types and sources of uncertainty, at least partly because they need to be treated in different ways in risk analyses.

In this chapter, we describe 2 approaches for classifying types of uncertainty and the hierarchical methods for propagating uncertainties that may be used with each classification scheme. We begin by providing an overview of the concepts of variability and uncertainty. Next, 2nd-order Monte Carlo is described because it is the technique most often used to propagate variability and uncertainty separately. The last section introduces a compatible method for dealing with uncertainty arising from incomplete data sets or partially relevant information: Bayesian hierarchical modeling. Simple case studies are provided to illustrate both techniques.

7.2   VARIABILITY AND UNCERTAINTY

There are many sources or components of uncertainty in an ecological risk assessment of a pesticide. For example, we may be uncertain about the identity of the species at highest risk of exposure, possible routes of exposure, the appropriate exposure model, ingestion rates, concentrations in different media, species sensitivity to the chemical of interest, and importance of modifying factors (e.g., proportion of time spent foraging in contaminated area). These sources of uncertainty generally belong to 1 of 4 general types of uncertainty: variability, uncertainty arising from lack of knowledge about parameter values, model structure, and decision rules. For a more in-depth discussion of these types of uncertainty, see Finkel (1990).

Variability refers to observed differences in a population or parameter attributable to true heterogeneity (Brusle 1991). It is the result of natural random or stochastic processes and stems from, for example, environmental, lifestyle, and genetic differences. Examples include variation between individuals in pesticide sensitivity and foraging behavior (e.g., time spent foraging in the agroenvironment) and between locations (e.g., soil type, climate, chemical concentration).

Parameter uncertainty refers to our uncertainty about the true values of the parameters or variables in a model (Smith and Shugart 1994). Parameters are typically estimated from laboratory, field, or other studies. This type of uncertainty is introduced because the estimated value may be based on insufficient, unreliable, or partially relevant information for the parameter of interest. Several processes contribute to parameter uncertainty including measurement errors, random errors, and systematic errors (Finkel 1990). Measurement error often arises from the imprecision of analytical devices used to, for example, quantify pesticide levels in different media. Errors in measurement, however, are not necessarily restricted to analytical hardware. Reconstructing pesticide use patterns in a region may be subject to measurement error because historical data can be faulty or ambiguous. Random error or sampling error is a common source of uncertainty in ecological risk assessment; it arises when one infers a quantity from a limited number of observations. For sample means, the importance of sampling error can be estimated by calculating the standard deviation (Sokal and Rohlf 1981). Sample means based on 3000 observations will have a standard deviation 1/10 that of means based on 30 observations. Systematic error occurs when the errors in the data are not truly random, such as might occur when the sample population is not representative of the entire population (e.g., when sampling is biased toward more contaminated areas in a crop field). Systematic error, unlike random error, does not decrease with more observations and is not accounted for when calculating sample statistics (e.g., mean, standard deviation). When systematic error is pervasive, sample statistics such as 95% confidence intervals can be quite misleading. For example, nearly half of the 27 measures of the speed of light measured between 1875 and 1958 had 95% or 99% confidence intervals that did not bracket the most accurate value available today ($c = 299\ 792.458$ km sec^{-1}) (Henrion and Fishhoff 1986).

In ecological risk assessment, we use mathematical models to determine which variables to measure, specify how they relate, and to estimate the values of variables we cannot measure directly. Model uncertainty is a serious challenge in ecological risk assessment (Finkel 1990; Reckhow 1994). Different dose–response models, for example, commonly lead to 2-fold or more differences in estimated low toxic effects doses (e.g., ED5 or LD10), even when the list of models is restricted to those that fit the data reasonably well ($p < 0.05$) and are theoretically plausible (Moore and Caux 1997). The problem of model uncertainty will be much more serious with complex models such as those used to estimate pesticide runoff, drift, and eventual fate in aquatic and terrestrial systems. Most applications of uncertainty analysis in ecological risk assessment do not propagate uncertainties associated with model structure, rather the model structure is assumed reasonable and only parameter uncertainties are propagated. Beck (1987), Reckhow (1994), Oreskes et al. (1994), and others discuss the issue of model uncertainty and describe the process for selecting, evaluating, calibrating, and validating models that, if followed, can substantially reduce this source of uncertainty in ecological risk assessment.

Decision rule uncertainty comes into play during risk management, i.e., after a risk estimate has been generated. This type of uncertainty arises when social objectives, economic costs, and value judgments are part of the decision-making process for determining which actions to take to remediate a problem. Individual decision makers are likely to be highly uncertain about how to best represent the complex preferences of their constituents. Such uncertainty can be quantified by collection of empirical data (e.g., opinion polls) and formally treated via decision analysis, but rarely is. Even with the availability of formal analytical tools, controversial judgments remain about how to value life, distribute costs, evaluate benefits and risks among individuals and groups, and decide whether to reduce risks now or some time in the future (Finkel 1990).

Most probabilistic assessments have tended to combine variability and parameter uncertainty, and not consider model or decision rule uncertainty. Recent guidance from the US National Academy of Sciences (NRC 1994), USEPA (1997), US DOE (Bechtel Jacobs Company 1998), and others (Hattis and Burmaster 1994; Hoffman and Hammonds 1994) has emphasized the importance of tracking variability and parameter uncertainty separately. Indeed, the USEPA (2000) states that "the risk assessor should strive to distinguish between variability and uncertainty." Two major advantages of tracking variability and parameter uncertainty separately in an uncertainty analysis are

- Precisely specifying the parameters of an input distribution is difficult and lacks credibility when the available information is limited. The alternative of using higher order techniques to represent uncertainty about the distribution parameters (see below) allows the assessor to be more forthright about what is known and what is not.
- The relative importance of variability and uncertainty can be quantified. This information can be used to determine whether further research would be useful (e.g., when uncertainty is much more important than variability) and to target that research. When variability is the dominant source of uncertainty, further research will be of limited use and the assessment should proceed to decision making.

The following section describes the most commonly used technique for propagating variability and parameter uncertainty separately, 2nd-order Monte Carlo analysis. A brief case study illustrating the technique is included in Section 7.3.

Second-order Monte Carlo analysis consists of 2 loops, the inner loop representing variability and the outer loop representing parameter uncertainty. To conduct an analysis, the following steps are required (also see Figure 7.1):

- Specify the model equation and identify which model inputs are 1) well-characterized constants (e.g., water solubility of a pesticide where there is little variation between a number of well conducted studies), 2) constants that have uncertainty (e.g., water solubility of a pesticide where only limited or poor quality data are available), 3) well-characterized random variables (e.g., pesticide concentration in a field from which numerous samples have been collected and analyzed), and 4) random variables for which there is uncertainty about the shape and/or parameters of the distribution (e.g., pesticide concentration in a field for which limited or poor quality data are available).

- In software systems such as Crystal Ball, well-characterized constants are assigned a single value that will be used in all simulations. Constants with uncertainty are assigned to the outer loop, and well-characterized random variables to the inner loop. For both situations, a distribution is selected (e.g., normal distribution for body weight) and the parameter values for the distribution specified (e.g., mean = 500 g, standard deviation = 100). For constants with uncertainty, the distribution and parameter values will likely be based on considerable professional judgment. Data-fitting techniques may be used to parameterize well-characterized random variables. Random variables with uncertainty must be included in both loops. To do this, a distribution is selected for the random variable (e.g., log-normal distribution for pesticide concentration) for the inner loop. Instead of specifying exact parameters for the random variable, however, distributions are assigned. In the case of a normal distribution in the inner loop, one would assign a distribution for the mean and/or a distribution for the standard deviation. These latter distributions would reflect our uncertainty about what the mean and/or standard deviation are for the random variable of interest.

- Specify the number of inner and outer loop simulations for the 2nd-order Monte Carlo analysis. In the 1st outer loop simulation, values for the parameters with uncertainty (either constants or random variables) are randomly selected from the outer loop distributions. These values are then used to specify the inner loop constants and random variable distributions. The analysis then proceeds for the number of simulations specified by the analyst for the inner loop. This is analogous to a 1st-order Monte Carlo analysis. The analysis then proceeds to the 2nd outer loop simulation and the process is repeated. When the number of outer loop simulations reaches the value specified by the analyst, the analysis is complete. The result is a distribution of distributions, a "meta-distribution" that expresses uncertainty both from uncertainty and from variability (Figure 7.1).

Inputs

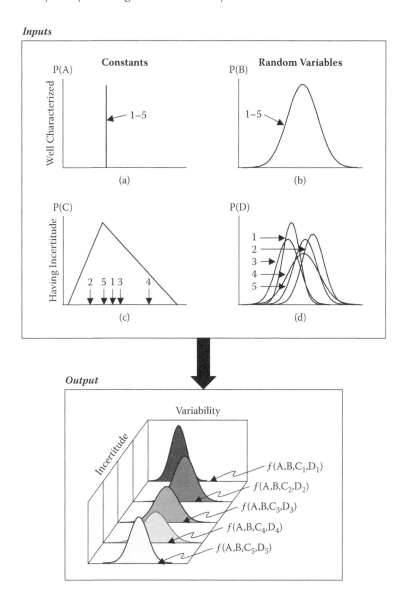

FIGURE 7.1   Use of 2nd-order Monte Carlo approach to distinguish between variability and uncertainty for mathematical expressions involving constants and random variables. Five hypothetical values or distributions from the outer loop simulation are shown for the inputs and output. For the well-characterized input constants and random variables, the values and distributions, respectively, do not change from one outer loop simulation to the next.

There are some issues associated with 2nd-order Monte Carlo analysis. Computational time can be a problem because the necessary number of replicates is squared with 2nd-order Monte Carlo analyses (i.e., number of inner loop simulations times number of outer loop simulations). In practice, specifying variability and uncertainty with random variables is a difficult exercise because the analyst is essentially trying to quantify what he or she does not know or only partially understands.

Issues involving dependencies become more complex in a 2nd-order Monte Carlo analysis (Hora 1996). As with 1st-order Monte Carlo analysis, dependencies can arise between different input variables (e.g., intake rates for air, water, and food) in 2nd-order Monte Carlo analysis. In 2nd-order Monte Carlo analysis, however, dependencies may also need to be specified between distribution parameters of a particular random variable. For example, means and standard deviations are typically correlated in nature; thus, for a normally distributed random variable, analysts must not only quantify what they do not know about the mean and standard deviation, but also what they do not know about the relationship between these parameters.

The major benefit of 2nd-order Monte Carlo analysis is that it allows analysts to propagate their uncertainty about distribution parameters in a probabilistic analysis. An analyst need not specify a precise estimate for an uncertain parameter value simply because one is needed to conduct the simulation. The relative importance of our inability to precisely specify values for constants or distributions for random variables can be determined by examining the spread of distributions in the output. If the spread is too wide to promote effective decision making, then additional research is required.

7.3   SIMPLE 2ND-ORDER MONTE CARLO ANALYSIS CASE STUDY

To illustrate the application of 2nd-order Monte Carlo analysis, we estimated exposure of Carolina wrens to a hypothetical pesticide in cotton fields in the southwest United States. For this case study, the pesticide is assumed to be persistent, and the goal is to estimate chronic exposure (i.e., total daily intake) at the local field scale. The input data are representative of the kinds of data available during reregistration but, for this case study, are entirely hypothetical.

The Carolina wren is primarily an insectivorous bird (94% of diet). It is nonmigratory and common in the cotton belt area of the southeastern United States. This species has been frequently observed foraging in or near cotton fields. Total daily intake of the hypothetical pesticide was estimated using the equation

$$TDI = \frac{IR_w\left[\left(C_{fw} \times P_f\right)+\left(C_{ew} \times P_e\right)\right]+IR_d\sum_{x=1}^{3}F_x\left[\left(C_{fx} + P_f\right)+\left(C_{ex} \times P_e\right)\right]}{BW} \quad (7.1)$$

where
TDI = Total daily intake (mg (kg·bw)$^{-1}$ day^{-1})
IR_w = Intake rate for water (L day^{-1})
IR_d = Dietary intake rate (kg day^{-1})

C_{fw} = Concentration in field water (mg L^{-1})
C_{ew} = Concentration in edge water (mg L^{-1})
C_{fx} = Concentration in field dietary items (mg kg^{-1}): plants ($x = 1$), insects ($x = 2$), and soil ($x = 3$)
C_{ex} = Concentration in edge dietary items (mg kg^{-1}): plants ($x = 1$), insects ($x = 2$), and soil ($x = 3$)
P_f = Proportion of time spent in the field
P_e = Proportion of time spent in the edge area (i.e., 50 m from field)
F_x = Fraction of diet for plants ($x = 1$), insects ($x = 2$), and soil ($x = 3$)
BW = Body weight (kg)

The state of the input knowledge can be summarized as follows (see also Table 7.1).

TABLE 7.1

Inputs for a 2nd-order Monte Carlo analysis to estimate exposure of Carolina wrens to a hypothetical pesticide (random variables are included in the inner loop of the Monte Carlo analysis, while random variables with uncertainty are included in both the inner and outer loops of the Monte Carlo analysis)

Input variable	Type of variable	Distribution and parameters
IR_w	Random variable	Log-normal, mean = 0.0041, SD = 0.001
IR_d	Random variable	Log-normal, mean = 0.0049, SD = 0.0009
C_{fw}	Random variable with uncertainty	Log-normal, mean = [normal, mean = 2.91, SD = 0.5], SD = [normal, mean = 0.34, SD = 0.05]
C_{ew}	Random variable	Log-normal, mean = 0.29, SD = 0.03
C_{f1} (plants)	Random variable with uncertainty	Log-normal, mean = [normal, mean = 0.18, SD = 0.03], SD = [normal, mean = 0.2, SD = 0.02], $r = 0.3$ (C_{f1}:C_{f3})
C_{f2} (insects)	Random variable with uncertainty	Log-normal, mean = [normal, mean = 2.92, SD = 0.5], SD = [normal, mean = 3.47, SD = 0.5]
C_{f3} (soil)	Random variable with uncertainty	Log-normal, mean = [normal, mean = 0.09, SD = 0.02], SD = [normal, mean = 0.05, SD = 0.01]
C_{e1} (plants)	Random variable	Log-normal, mean = 0.0018, SD = 0.002, $r = 0.35$ (C_{e1}:C_{e3})
C_{e2} (insects)	Random variable	Log-normal, mean = 0.04, SD = 0.04
C_{e3} (soil)	Random variable	Log-normal, mean = 0.0009, SD = 0.0009
P_f	Random variable with uncertainty	Beta, alpha = [normal, mean = 3, SD = 1], beta = [normal, mean = 3, SD = 1], scale = 0.6, $r = -0.9$ (P_f:P_e)
P_e	Random variable with uncertainty	Beta, alpha = [normal, mean = 5.2, SD = 1], beta = [normal, mean = 3, SD = 1], scale = 1
F_1 (plants)	Constant	0.94
F_2 (insects)	Constant	0.04
F_3 (soil)	Constant	0.02
BW	Random variable	Normal, mean = 0.0186, SD = 0.0019

1) The diet exhibits little variability and is well characterized. In this case study, the dietary fractions for insects, plants, and soil are treated as constants.
2) Body weight is a well-characterized random variable.
3) Some data are available from several fields and field edges to estimate spatial and temporal variability around 30-day mean concentrations. In this case study, the concentration variables for the field are treated as random variables with uncertainty. Because the concentration variables for the edge are minor contributors to total daily intake, the uncertainty about these variables is ignored.
4) The time individuals spend foraging in fields and field edges is expected to be variable, but data are scant. The foraging behavior variables are treated as random variables with large uncertainty.
5) The intake rates are based on well-characterized allometric relationships. In this example, these variables are treated as random variables without uncertainty.

Table 7.1 shows the input variables derived from the above state of knowledge for the 2nd-order Monte Carlo analysis. Correlations between pesticide concentrations in soil and plants in the field and edge were also specified for this analysis, as was a strong negative correlation between the variable, proportion time spent in the field, and proportion time spent in the edge habitat. For simplicity, no correlations were specified between the 2nd-order input parameters (e.g., 2nd-order mean and standard deviation for concentration in insects in the field, 2nd-order alpha and beta for proportion of time spent in the field) even though they are unlikely to be independent. The analysis was run with 500 inner loop simulations and 250 outer loop simulations ($500 \times 250 = 125,000$ simulations) using Latin Hypercube sampling.

The results of the 2nd-order Monte Carlo analysis are shown in Figure 7.2. Rather than show the output distributions for all 250 outer loop simulations, the figure shows the 5th, 25th, 50th, 75th, and 95th percentile distributions for total daily intake (TDI) of the hypothetical pesticide by Carolina wrens. The results indicate that, because of uncertainty about some of the input parameter distributions, there is a 5% probability that the median TDI is less than 0.224 mg/kg bw/day and a 95% probability that the median TDI is less than 0.557 mg/kg bw/day (in Figure 7.2, read over from the cumulative probability of 50% on the y-axis to the 5th percentile and 95th percentile output distributions). Similarly, there is a 5% probability that the 10th percentile TDI is less than 0.103 mg/kg bw/day and a 95% probability that the 10th percentile TDI is less than 0.313 mg/kg bw/day. Finally, there is a 5% probability that the 90th percentile TDI is less than 0.484 mg/kg bw/day and a 95% probability that the median TDI is less than 1.09 mg (kg bw)$^{-1}$ day^{-1}. If the corresponding effects benchmark or distribution is well left of the 5th percentile output distribution or well right of the 95th percentile output distribution, then the uncertainty about the input distributions is inconsequential. If, however, the effects benchmark or distribution lies between the 5th and 95th percentile output distributions, then additional research effort may be needed to reduce uncertainty about the TDI distribution.

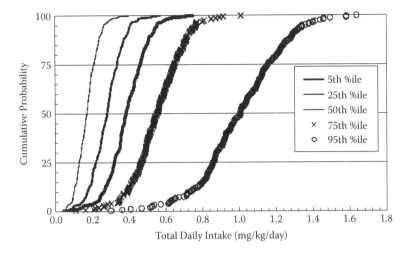

FIGURE 7.2   Total daily intake for Carolina wrens exposed to a hypothetical pesticide.

7.4   BAYESIAN HIERARCHICAL MODELING

In many risk assessments, assessors are faced with the challenge of incomplete data sets, a small number of tests on the chemical of interest, inconsistent information between studies, and other issues that lend uncertainty to the risk assessment results. Bayesian hierarchical models provide a tool for minimizing the effect of these issues (see Gelman et al. 1998 for a discussion of Bayesian methods and Bayesian hierarchical models in particular). For example, the Bayesian concept of exchangeability (Gelman et al. 1998) effectively minimizes the loss of information associated with small sample sizes by allowing the exchange of information among samples. A model based on small data sets can effectively "learn" from data conducted on similar tests (i.e., exchangeable tests), thus effectively maximizing the information content in the entire data set. Also, information can be output at each hierarchical level, thus communicating information at various levels of aggregation. Moreover, the methods enable direct incorporation of subjective information through the use of Bayes' theory (see Chapter 5 of this book). In this section, we illustrate the use of Bayesian hierarchical models as an alternative to 2nd-order Monte Carlo or other higher order methods for tackling problems with high uncertainty. The concept of Bayesian hierarchical modeling is illustrated for the problem of estimating species sensitivity distributions (SSDs).

SSDs are being routinely used for the display and interpretation of effects data (Parkhurst et al. 1996; Posthuma et al. 2002). An SSD for atrazine (shown in Figure 7.3) displays the typical S-shaped curve associated with many chemical dose–response relationships. Each point on the curve represents an LC50 for a particular species exposed to atrazine under standard toxicity test protocols. The SSD approach uses only a single statistically derived endpoint from each available toxicity test (e.g., the LC50 or EC50). In contrast, all data collected during any specific toxicity test can be used in a hierarchical model. The ability to use all available data to make inferential decisions is a marked improvement over the standard SSD effects distribution.

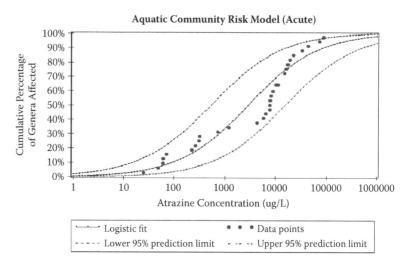

FIGURE 7.3   Species sensitivity distribution for atrazine.

7.4.1 CASE STUDY EXAMPLE OF BAYESIAN HIERARCHICAL MODELING

Sixteen nontarget plant trials using various study designs were conducted with a herbicide, its degradates, and formulations under growth chamber and greenhouse conditions. Each study or trial included tests with multiple species. Among the 16 studies, a total of 17 species were tested, some of them tested in multiple studies. Where multiple tests on a single species were available, the data were combined into a single dose–response analysis as described in Section 7.4.1.2.

Shoot weight from vegetative vigor studies was chosen as the primary measurement endpoint to assess the risk to standing vegetation. Shoot weight is more reliable than root weight and clearly more sensitive than shoot length under the vegetative vigor study design. Shoot length data from seedling emergence studies were selected as a measurement endpoint to assess the risk to emerging vegetation. In both study designs, mortality was too infrequent to use as a measurement endpoint. Vegetative vigor shoot weight and seedling emergence shoot length are sensitive and reliable measures of phytotoxicity and are widely accepted in the regulatory and scientific community (Davy 2001).

Seventeen species (42 tests) were tested with the vegetative vigor study design: alfalfa, beets, cabbage, canola, corn, cotton, cucumber, lettuce, navy bean, oat, onion, radish, ryegrass, soybean, sunflower, tomato, and turnip. Ten species (25 tests) were evaluated using the seedling emergence study design: cabbage, corn, cucumber, lettuce, oat, onion, ryegrass, soybean, tomato, and turnip.

7.4.1.1 Bayesian Inference

Two very different approaches to inferential statistics exist: the "classical" or "frequentist" approach and the Bayesian approach. Each approach is used to draw conclusions (or inferences) regarding the magnitude of some unknown quantity, such as the intercept and slope of a dose–response model. The key difference between classical and Bayesian statistics lies in the concept of probability used by each approach. In the classical approach, probability represents the frequency with which an event would occur in repeated trials. In Bayesian statistics, probability represents a degree of reasonable belief based on existing information. Bayesian inference assumes that the parameters of interest are random. Available data are used to make inferences (e.g., probability statements) about the random parameters. This information assumes 2 forms: sample information and prior information. Each must be available for the Bayesian paradigm to be implemented. Probability statements about the parameters of interest are made based solely on these 2 sets of information. Sample information and prior information are combined through the equation underlying Bayes's theory. The final product of the Bayesian procedure is a probability distribution (called the posterior distribution) of the random parameter. Area under the curve of the posterior distribution is used to make probability statements about the random variable. These areas are called credible intervals to reflect the concept of probability represented by the distribution.

In this section, the sample information was the raw toxicity test data for each species and test. Prior information was not available outside the data set, so vague prior information was used as a basis for implementing the procedures.

A major advantage of the Bayesian framework for this risk assessment was the ability to make probability statements across a hierarchy of data levels. Probability distributions of the random parameters could be easily combined (integrated) across data levels. For example, information could be combined across tests within a laboratory or across laboratories to make inferences about the random parameters of interest. The hierarchical model used in this project is described in Section 7.4.1.3. Software systems such as WinBugs (Spiegelhalter et al. 2000) facilitate the process. The advent of the new software reduces the level of mathematical and programming skill historically required to implement the Bayesian paradigm. For more on Bayes's theory and decision-theoretic approaches, see Gelmen et al. (1998) and Congdon (2001).

7.4.1.2   Development of Dose–Response Functions for Individual Species

The effects information was derived from toxicity tests in which replicate pots containing multiple plants were exposed to a range of herbicide doses. Exposure in the toxicity tests was expressed as mass of active ingredient (a.i.) applied per horizontal surface area (lb a.i. A^{-1} or equivalent g a.i. ha^{-1}). Pots, or groups of pots, were treated as replicates under the study designs used. The measured test endpoint used in the effects assessment for each replicate was shoot weight for vegetative vigor studies or shoot length for seedling emergence tests. The replicate values were used in all statistical analyses.

A number of issues influenced the selection of the dose–response model form and the treatment of the data prior to fitting the model. First, shoot weight and shoot length are continuous response measurements; therefore, use of a standardized logistic model form is not appropriate. Second, the natural variation in plant growth often resulted in apparent increased shoot weight and shoot length measurements relative to the control at low herbicide application rates. A dose–response model needs to perform well even when some measurements in treatment levels exceed the controls.

Third, between-test variability can be attributed to many conditions, including differences in soil and nutrient conditions, as evidenced by differences in control performance. Therefore, a method was needed to normalize the toxicity test results so they could be combined across tests. Finally, the physical nature of the test ensures that 100% inhibition can never be reached, because a plant experiencing 100% inhibition would be counted as a mortality. As a result, no measured data points were available near the 100% inhibition level. For most tests, the greatest measurable response was 70% to 80% inhibition. Some standard dose–response models will not fit the resulting test data well, and will result in biased estimators of the true EC25 and EC50 values for specific tests. The following paragraphs describe the approaches that were selected to resolve these issues.

A standard approach was used to model the data after normalization relative to the appropriate control. The raw shoot weight or shoot length data were normalized by dividing by the control mean for each test to transform the endpoint to the fraction of the control (which was used as the response endpoint in the model). This transformation is presented as:

$$y_{i,j,k,l} = \frac{\text{dose}_{i,j,k,l} \text{ response}}{\text{control mean response}_{j,l}} \tag{7.2}$$

where i is the index of individual replicates, j is the index of test, k is the index of dose level, and l is the index of species. Each test had from 3 to 10 replicates at each dose. The control mean response for each test was calculated using values from all control replicates in the test. When there were multiple tests, shoot weight or shoot length data from each test were rescaled by their respective control means to adjust for possible differences in test conditions between studies. The rescaled data were modeled as a function of the dose level:

$$y_{i,j,k,l} = \left(R_{0,l} - W_{0,l}\right) \frac{\exp\left(\alpha_{l,j} + \beta_{l,j} \log\left(C_{j,k,l}\right)\right)}{1 + \exp\left(\alpha_{l,j} + \beta_{l,j} \log\left(C_{j,k,l}\right)\right)} + W_{0,l} + \varepsilon \tag{7.3}$$

$R_{0,l}$ is the rescaled response at the control, $C_{j,k,l}$ is the dose or rate, and ε is the error term, and α and β are regression coefficients. $W_{0,l}$ is the expected lowest shoot weight (or length) value, and effectively rescales the model predictions over the dose–response range. Both $R_{0,l}$ and $W_{0,l}$ were calculated for each species-specific dose–response curve.

7.4.1.3   Bayesian Hierarchical Model

A Bayesian hierarchical modeling framework was used to evaluate the effects data for each species and test endpoint (Figure 7.4). Hierarchical models reduce the effect of incomplete data sets, small numbers of tests, inconsistent information on effects among species, and other issues that lend uncertainty to the risk characterization results.

Dose–response models were fit individually for each species. Sources of uncertainty inherent in these models include differences among individual plants used

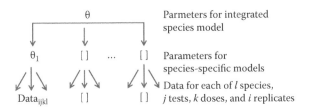

FIGURE 7.4 Bayesian hierarchical framework.

in the tests, differences among species responses, and the uncertainty about model parameters. In the hierarchical framework, differences among species can be treated as the result of another "super" distribution. The dose–response curves for individual species can be treated as samples from a distribution at a higher level, with each individual dose–response curve representing a random realization from this super distribution.

To summarize, let θ be the parameters R_0, W_0, α, and β; and π be a probability density function. The distribution π describes the variability of the model parameters. The objective of a Bayesian hierarchical model is to generate the distributions of these parameters, based on all available information.

The mathematics underlying the hierarchical model (Figure 7.4) are described in the following discussion.

The hierarchical model can be described in 4 levels. For convenience, the model in Equation (7.3) is expressed in short form as a function of unknown coefficients and the dose concentrations: $f(\theta_{j,l}, C_{i,j,k})$. The unknown coefficients

$$\theta_{jl} = \left(R_{0,jl}, W_{0,jl}, \alpha_{jl}, \beta_{jl} \right) \tag{7.4}$$

and the dose concentrations, $C_{i,j,k}$ define the mean of individual data points at the 1st level:

$$y_{i,j,k,l} \sim N\left(f\left(\theta_{jl}, C_{i,j,k} \right), \sigma^2 \right) \tag{7.5}$$

Intuitively, we assume that individual data points are generated by a normal distribution. A common variance is assumed for model error as in a conventional regression analysis.

At the second level, coefficients $\theta_{j,l}$ are modeled as random variables from species-specific distributions:

$$\theta_{jl} \sim N\left(\theta_l, \tau^2 \right) \tag{7.6}$$

where $\theta_l = (R_{0,l}, W_{0,l}, \alpha_l, \beta_l)$, representing model coefficients at the species level. The variance parameter τ^2 represents the within-species variance in model coefficients.

The species-level coefficients are further modeled as random variables from a common hyperdistribution at the 3rd level:

$$\theta_l \sim N\left(\theta, \lambda^2\right) \tag{7.7}$$

where $\theta = (R_0, W_0, \alpha, \beta)$ is the hyperparameter vector representing the overall means of model coefficients, and the variance λ^2 is the between-species variance of the model coefficients.

In the 4th level, prior distributions are defined for the variance parameters and the hyperparameter:

$$\theta \sim N\left(0, 0.0001\right)$$

$$\frac{1}{\lambda^2} \sim gamma\left(0.001, 0.001\right)$$

$$\frac{1}{\tau^2} \sim gamma\left(0.001, 0.001\right) \tag{7.8}$$

$$\frac{1}{\sigma^2} \sim gamma\left(0.001, 0.001\right)$$

Under this hierarchical model, the joint posterior distribution of all coefficients and parameters can be expressed as the product of the probability density functions at the 4 levels:

$$\pi\left(\theta_{j,l}, \theta_l, \theta, \sigma^2, \tau^2, \lambda^2 \mid Y\right) = \pi\left(Y \mid \theta_{j,l}, \sigma^2\right) \times$$

$$\pi\left(\theta_{j,l} \mid \theta_l, \tau^2\right) \pi\left(\theta_l \mid \lambda^2, \theta\right) \pi\left(\theta\right) \pi\left(\sigma^2\right) \pi\left(\tau^2\right) \pi\left(\lambda^2\right) \tag{7.9}$$

From this joint distribution, it is possible to integrate out coefficients and parameters at selected levels to summarize information at a given level. For example, the distribution $\iint \pi(\theta_l \mid \lambda^2, \theta) \pi(\theta) \pi(\lambda^2) d\lambda d\theta$ summarizes the behavior of model coefficients for species l, which can be used to summarize possible test-level outcomes. Likewise, when all other coefficients are integrated out of the joint distribution, the posterior distribution of $\pi(\theta \mid Y)$ represents information from all species.

Equations (7.5) to (7.8) represent the general structure of the model. Equation (7.9) is a mathematical picture for the entire framework and indicates how numerical integration can be used to generate parameter distributions at any level. At any point in the framework, we can calculate the EC25 and EC50 as

$$\log (EC50) = -\alpha/\beta \text{ and } \log (EC25) = [-\alpha/\beta - \log(3)/\_] \tag{7.10}$$

Output from the hierarchical model can be produced at any level. For this analysis, these outputs can be presented in both graphical and tabular form. For example, the median value of the model parameters for each species can be output and used to create graphics of the resulting dose–response curves that are generated using these coefficients. Also at this level, test endpoints like the EC25 for each pair of model coefficients generated during the statistical estimation process (see above equations) can be calculated. Because the Bayesian software uses a random sampling procedure to solve the Bayes rule, many sets of model parameters for each species-specific dose–response model are produced. For each randomly generated set of parameters, the corresponding EC25 and EC50 is calculated. These values are random samples from the posterior distribution of the EC25 or EC50. At the various hierarchical levels, these values provide information that can be statistically evaluated or plotted.

For example, a cumulative distribution of the EC25 values integrated across all species (for shoot length or shoot weight) can be readily created. While this distribution is derived based on the available data and model parameters, it represents the entire range of EC25 values that could be encountered. The distribution represents among-species variability and incorporates all associated sources of variance, including species sensitivity, model fit, and random error. From the generated EC25 values, a cumulative distribution of EC25 values that is consistent with the concept of SSDs currently in the literature can be developed. The advantage of the Bayesian hierarchical model approach is that the resulting distribution of EC25 values incorporates many sources of variability, without loss of information.

The WinBugs software system (Spiegelhalter et al. 2000) was used to solve the Bayesian framework equations. These solutions result in posterior distributions of the random parameters that are effectively 1) the model parameters for the dose–response models at both the among-species and superpopulation levels, 2) used to calculate measures of variability in the EC25s and EC50s, and 3) used to calculate the expected values of the model parameters at each level. WinBugs uses Markov Chain Monte Carlo (MCMC) techniques to solve the integrals found in Bayes's theorem, conditional on the distributional form of the parameters. To run WinBugs, the user supplies 1) the model form, 2) the distributional form of all random parameters at each level of the hierarchical model, 3) prior distributions of the parameters, and 4) any calculations involving the random parameters that the user wants the computer to generate. The software system uses random sampling of the conditional distributions to solve Bayes's theorem, resulting in the posterior distribution of the random parameters, conditional on the data. The user can output sufficient statistics of the random parameters at any point in the model hierarchy. Details on the MCMC approach for solving the Bayesian equations are given in Congdon (2001).

7.4.1.3   Species Sensitivity Distributions

A species sensitivity distribution (SSD) is a statistical distribution describing the variation in toxicity among a set of species. The set of species can be composed of a specific taxon, a selected species assemblage, or a natural community. Because the true interspecies distribution of toxicity is unknown, the SSD is generated from existing toxicity information and is presented as a cumulative distribution function. A basic assumption in ecological risk assessment is that laboratory-generated single-species toxicity data provide useful information about the communities to be protected. In practice, however, not all species can be tested due to experimental or financial limitations. The SSD provides a statistical tool for extrapolating from limited data sets to other species and situations not directly available from existing toxicity test information.

Bayesian models were used to generate the EC25 values for use in the SSDs because the Bayesian approach generates the most complete expectation of EC25 values that can be developed from the available information. The Bayesian-derived SSD represents the entire range of EC25 values that could be encountered. Standard SSDs depend only on the specific species EC25 (or EC50) values available, reflecting interspecies variability for a fixed number of species. Uncertainty in the species-specific EC25 values is not represented in standard SSDs. The Bayesian-derived SSD, however, is influenced not only by the species in the effects data set, but also by the uncertainty due to random error, intraspecies and interspecies variation, and model fit. Incorporation of additional sources of variability into the SSD is a major advantage of the Bayesian hierarchical framework relative to the classical approach. The Bayesian framework begins with the raw data and its inherent variability and uses that variability to calculate the EC25 values for each test and endpoint. From the generated EC25 values, a cumulative distribution of EC25 values is developed that is consistent with the concept of SSDs currently in the literature (Posthuma et al. 2002). The Bayesian hierarchical approach therefore results in a complete representation of the information inherent in the toxicity database.

7.4.1.4   Results

Examples of the model fit for shoot weight and shoot length are shown in Figures 7.5 and 7.6, respectively. The models fit the data well, particularly at the low concentrations where risk-based decisions are typically focused. The benefit of the additional model parameter, $W_{0,1}$, is evident by the "floor" effect seen in the raw data at higher concentrations. Both data-rich and relatively data-poor data sets followed the shape of the model curve.

Figures 7.7 and 7.8 show the dose–response curves generated at the species level for shoot weight and shoot length, respectively. Each dotted curve on the plot is species specific. For many species, multiple toxicity tests are available. The random

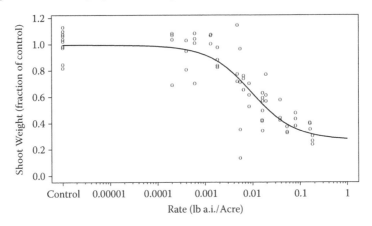

FIGURE 7.5   Cucumber shoot weight dose–response data and model.

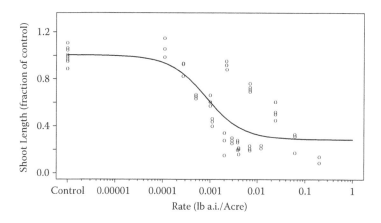

FIGURE 7.6   Turnip shoot length dose–response data and model.

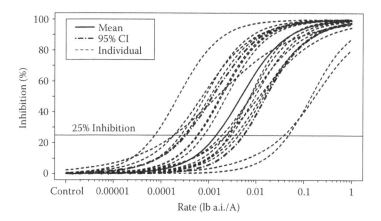

FIGURE 7.7   Integrated Bayesian effects for shoot weight.

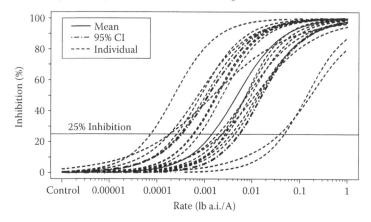

FIGURE 7.8   Integrated Bayesian effects plots for shoot length.

parameters in the species-specific dose–response model are generated by integrating their respective probability distributions across the test-specific parameter distributions. Using Equation (7.10), the MCMC sampling techniques provide random samples of the EC25 and EC50 that reflect the scale (the center) and shape (the variance) of the parameter distributions. Therefore, the shape and spread of the species-specific distributions EC25 and EC50 distributions shown in Figures 7.7 and 7.8 reflect the relative scale and shape of the parameters that are lower in the hierarchical model framework. The species-specific parameter distributions effectively reflect several sources of variation including model error and between-test variability.

The solid line in the center of Figures 7.7 and 7.8 represents the integrated effects model, reflecting the average expected effects for all species based on the available shoot weight information. The darker dotted lines on either side are the 95% credible intervals around the integrated effects curve, and the dotted lines are the individual species-specific dose–response curves. Although not used in this risk assessment, the upper 95% credible interval provides a conservative measure of risk. It could be interpreted as an upper bound or a hypothetical species that has a 95% chance of occurring (although not directly measured). The solid horizontal line indicates the EC25 values. The graphic, which is easily output from the Bayesian hierarchical procedures, provides a clear perspective on the among-species variation in sensitivity.

The advantages of this method are 1) the approach provides a visual interpretation of the relative effects of each species on the integrated model, 2) all of the raw data can be used to develop the integrated curve, 3) specific test endpoints (e.g., EC25) can be taken from the curves for each species (either mathematically or visually), resulting in a data representation similar to the SSD, and 4) a formal mathematical representation of the curves and probability model is available, unlike the SSD approach. Visually, an SSD-type representation can be seen by first picking an effects level on the *y*-axis (inhibition) and moving horizontally across the graph. The points on each dose–response curve are those typically included in an SSD.

7.5 REFERENCES

Bechtel Jacobs Company. 1998. Guidance for treatment of variability and uncertainty in ecological risk assessments of contaminated sites. Oak Ridge (TN): US Department of Energy. BJC/OR-55.

Beck MB. 1987. Water quality modeling: a review of the analysis of uncertainty. Water Resour Res 23:1393–1442.

Brusle J. 1991. The eel (*Anguilla* sp.) and organic chemical pollutants. Sci Total Environ 102:1–19.

Congdon P. 2001. Bayesian statistical modelling. New York: Wiley.

Davy M. 2001. Terrestrial plant tests and evaluation. In: FIFRA Scientific Advisory Panel Open Meeting June 27 to 29, 2001: review of non-target plant toxicity tests under NAFTA. Washington (DC): USEPA, OSCP.

Finkel, A. M. 1990. Confronting uncertainty in risk management. a guide for decision-makers. Washington (DC): Center for Risk Management, Resources for the Future.

Gelmen A, Carlin JB, Stern HS, Rubin DB. 1998. Bayesian data analysis. London: Chapman and Hall.

Hattis D, Burmaster DE. 1994. Assessment of variability and uncertainty distributions for practical risk analyses. Risk Analysis 14:713–730.

Henrion M, Fishoff B. 1986. Assessing uncertainty in physical constants. Am J Phys 54:791–797.

Hoffman FO, Hammonds JS. 1994. Propagation of uncertainty in risk assessments: the need to distinguish between uncertainty due to lack of knowledge and uncertainty due to variability. Risk Anal 14:707–712.

Hora SC. 1996. Aleatory and epistemic uncertainty in probability elicitation with an example from hazardous waste management. Reliability Eng Syst Saf 54:217–223.

Ludwig D. 1996. Uncertainty and the assessment of extinction probabilities. Ecol Appl 6:1067–1076.

McNeill D, Freiberger P. 1993. Fuzzy logic. New York: Simon and Schuster.

Moore DRJ, Caux P-Y. 1997. Estimating low toxic effects. Environ Toxicol Chem 16:794–801.

Morgan MG, Henrion M. 1990. Uncertainty: A guide to dealing with uncertainty in quantitative risk and policy analysis. Cambridge (UK): Cambridge University Press.

[NRC] National Research Council. 1994. Science and judgment in risk assessment. Committee on Risk Assessment and Hazardous Air Pollutants. Washington (DC): National Academy Press.

Oreskes N, Shrader-Frechette K, Belitz K. 1994. Verification, validation, and confirmation of numerical models in earth sciences. Nature 263:641–646.

Parkhurst BR, Warren-Hicks WJ, Cardwell RD, Volosin J, Etchison T, Butcher JB, Covington SM. 1996. Methodology for aquatic ecological risk assessment. Alexandria (VA): Water Environment Research Foundation.

Posthuma L, Suter GW II, Traas TP, editors, 2002. Species sensitivity distributions in ecotoxicology. Boca Raton (FL): Lewis Publishers.

Reckhow KH. 1994. Water quality simulation modeling and uncertainty analysis for risk assessment and decision making. Ecol Model 72: 1–20.

Rowe WD. 1994. Understanding uncertainty. Risk Anal 14:743–750.

Smith EP, Shugart HH. 1994. Uncertainty in ecological risk assessment. In: Ecological risk assessment issue papers. Washington (DC): Risk Assessment Forum, US Environmental Protection Agency. USEPA 630/R/94/009.

Sokal RR, Rohlf FJ. 1981. Biometry. New York: W.H. Freeman and Company.

Spiegelhalter D, Thomas A, Best N. 2000. WinBugs version 1.3 user manual. Available from: http://www.nrc-bsu.cam.ac.uk/bugs

[USEPA] US Environmental Protection Agency. 1997. Guiding principles for Monte Carlo analysis. Office of Research and Development. Washington (DC): US Environmental Protection Agency. USEPA/630/R-97/001.

[USEPA] US Environmental Protection Agency. 2000. Risk characterization handbook. Washington (DC): Office of Research and Development, US Environmental Protection Agency. USEPA 100-B-00-002.

8 Interpreting and Communicating Risk and Uncertainty for Decision Making

J. L. Shaw, K. R. Tucker, K. Corsten,
J. M. Giddings, D. M. Keehner, and C. Kriz

8.1 INTRODUCTION

Effective communication of ecological risk assessment results, and the uncertainty around these "predictions" of ecological risk, is essential for producing the best possible risk management decisions and for ensuring a quality dialogue among all stakeholders regarding risk management options. The complexity and unfamiliarity of probabilistic methods and the nature of the results for these audiences presents a particular challenge to risk assessors. They must not only master the analysis and accurately interpret results but be able to communicate results and process in precise nontechnical language that adequately describes uncertainty, as well as science policy options, and choices made along the way. Ambiguous, inexplicit, or inaccurate interpretations of science policy choices and risk assessment results contribute to poor decision making and an inefficient use of societal resources. In this chapter, we provide guidance for practitioners of risk assessment, particularly risk assessors and decision makers on effective communication throughout the ecological risk assessment process. Good communication among practitioners is not only essential for informed regulatory decisions on pesticides, but lays the groundwork for effective communication with stakeholders and ultimately with the public about risk assessment results and risk management options and decisions.

Many differences exist among regulatory agencies in different countries. For ease of presentation, we generalize our discussion based on the separation of assessor and manager or decision-maker roles in the United States while recognizing that they may be dealt with in other countries via a more consensus-based system that combines assessor and risk manager roles. Other differences exist where some regulatory agencies require a balancing of ecological risk with pesticide benefits (in terms of impacts on crop yield and quality) prior to making regulatory decisions, while others may simply require that the decision maker be aware of the pesticide's benefits to crop production. Regardless of the various legal and statutory frameworks for pesticide regulation across and within countries, or the designated roles and responsibilities of the people involved, clear interpretation and communication of risk assessment methods and results by experts remains an essential part of effective decision making.

8.2   PARTICIPANTS IN RISK COMMUNICATION

Risk communication is defined as an interactive process of risk information and opinion among individuals, groups, and institutions (NRC 1989). The process is critical to effective decision making and to the exchange of accurate information within and between several, often overlapping, categories of participants. Table 8.1 provides an example of participants involved in this process and their roles. For the purposes of this chapter, we use the term practitioner to refer to the regulators who are responsible for initiating, leading, and implementing ecological probabilistic risk assessments for pesticide registration, and those registrants who develop ecological risk assessments. Most narrowly, practitioners engage in 2 interrelated activities: assessment and decision making. The latter role may be filled by risk assessors, risk managers, or some combination of these or other personnel.

Stakeholders may include individuals with a variety of affiliations or personal interests, including academic institutions, nonprofit organizations, other government agencies not directly engaged in the practice of pesticide registration, and the public. Stakeholders feel a mutual responsibility for the nature of any assessment outcome and a need to share resources and information to ensure a fully informed decision. For the most part, the public and many NGOs, government agencies, trade groups, public interest groups, and others remain removed from the implementation of the process of probabilistic risk assessment itself and focus more on the anticipated outcomes and effects of the decision. However, they are participants and potential stakeholders because their views or interests are represented in theory by the stakeholders, particularly government decision makers who are mandated to represent the citizenry they serve, and yet they may become invested economically, professionally, or personally and choose to take a more active role in the process.

Participants, stakeholders, and practitioners interact in the stakeholder process. Each group is inclusive of the other, and each group is responsible for applying, interpreting, and reviewing the ecological assessments and the uncertainty analysis methods where they are used.

8.3   COMMUNICATING UNCERTAINTY TO
STAKEHOLDERS AND PARTICIPANTS

There are many reasons why communicating risk and uncertainty to stakeholders and participants is critical to an informed assessment but 3 are, perhaps, most fundamental. First, participants (especially stakeholders with expertise in topics germane to probabilistic risk assessment or a particular assessment), if given opportunities to interact with practitioners, can contribute information and perspectives that could help focus and

Table 8.1   An example of participants that could be involved in the risk communication process for pesticides

Type of participant	Role in process	Communication
Regulator–risk assessors (directly involved or peer-reviewers)	Preparation of a scientifically valid risk assessment	Technical experts with responsibility for communicating results to decision makers and other stakeholders
Regulator or registrant–benefits assessors	Responsible for benefits evaluation (where the law requires this)	Technical experts with responsibility for communicating results to others
Regulator–decision makers	Compliance with laws regulating pesticides; consideration of input from all stakeholders; makes decision to register or reregister the pesticide based on risk assessment and benefits analyses; makes decision on mitigation measures necessary to support registration decision	Assist in communication to all nonregulator stakeholders — understanding of risk assessment necessary
Registrant–risk assessors (directly involved or peer-reviewers)	Preparation of a scientifically valid risk assessment	Technical experts with responsibility for communicating results to regulators, registrant–decision makers, and others
Registrant–decision makers	Makes decision to develop the pesticide and to submit pesticide for registration based on risk assessment and benefits analyses conducted by the registrant; makes decision on mitigation measures	Assists in communication to regulators, and other stakeholders — understanding of risk assessment necessary
Regulators at a local level or public authorities	Compliance with laws at local level	Recipients for communication who may have limited knowledge of risk assessment
Food, feed, or fiber processors	Focus on potential benefits and need for low-risk pesticide	Recipients of communication with limited knowledge of risk assessment
Universities or researchers	May be consulted in decision making	Recipients of communication who may have limited knowledge of risk assessment assuming they are not practitioners of risk assessment

—continued

Table 8.1   (continued) An example of participants that could be involved in the risk communication process for pesticides

Type of participant	Role in process	Communication
Legislators	Risk communication may influence and facilitate the legislative process	Recipients of communication with limited knowledge of risk assessment but need to understand the role of science in informing policy
Consumer protection groups	Represent consumer rights to know	Recipients of communication with limited knowledge of risk assessment
Environmental groups	Represent perspectives and interests of environmental groups	Represents perspectives and interests of environmental groups
Public	Individual rights and interests	Recipients of communication with limited knowledge of risk assessment and focus on perceived risk; perceptions of risk based on personal experiences
Resource managers with mandate to protect resources	Represent interests of natural resources; may be consulted in problem formulation and decision making	Recipients of communication with limited knowledge of risk assessment

refine the assessment. Second, effective communication of uncertainty gives participants an appreciation of the limits of the data and of the scientific tools employed. This transparency can build trust and help dispel fears that the scientific findings regarding risk are being misrepresented. Third, greater awareness of the treatment of uncertainty in the assessment, and the limits of the science, enlightens decision makers when making policy choices in the face of uncertainty. That is, with increased transparency and clarity around the limits of our best estimates of risk, it will become increasingly clear that science informs but does not dictate risk management decisions.

Regulatory agencies bear the primary responsibility for risk communication to stakeholders and participants, including the public. Regulators and registrants are most involved in initiating risk communication. Other federal agencies, NGOs, interest groups, and regional governments may disseminate information geared to stakeholders and the public that more often discusses risk in specific use scenarios and, in some cases, is deliberately biased or "protective" of a particular position or species deemed to be at risk.

The NRC (1989) states "risk communication is successful to the extent it raises the level of understanding of relevant issues or actions and satisfies those involved that they are adequately informed within the limits of available knowledge." Central to many of these discussions is the need for participants to feel comfortable with the amount of information available and know they have an opportunity to use this information constructively.

Three generalized tasks contribute to these elements of successful risk communication. First, the goal of risk communication is not persuasion or simply delivering "the message," but rather it is to provide the resources, information, and expertise that enable participants to make an informed decision. In the case of some participants who become only indirectly involved in the process, education on fundamental concepts may be necessary (Peterson 2000). Powell and Leiss (1997) warn that educating the public about science is no substitute for good risk communication practice. In response to this sound advice, where probabilistic risk assessment is just entering registration practice, regulators may consider approaches that creatively include educational opportunities (developing education or training on risk assessment) in laying the groundwork for effective communication.

Third, whenever possible, this communication needs to be interactive. In the registration of pesticides, legally mandated deadlines, the complexity of the assessment, the manifold assumptions needed to fully explore methods and their outputs, and the need to protect sensitive business information limits the degree to which stakeholders may always be able to participate interactively. Nevertheless, opportunities for meaningful interaction with stakeholders, especially those that can bring additional data or information to bear and can constructively review the process, will benefit practitioners.

8.4   PROCESS FOR COMMUNICATION

Risk communication is defined as an interactive process of risk information and opinion among individuals, groups, and institutions (NRC 1989). In other words, risk communication is not simply about experts communicating the results of a risk assessment following its completion. Instead, this interactive process requires a dialogue that should begin in the problem formulation stage. Problem formulation is the 1st formal stage of communication and decision making and is the starting point for effective dialogue throughout the remainder of the process. An inadequate problem formulation will hinder communicating results of the risk characterization. Also, risk assessments that do not take advantage of the concerns and expertise of stakeholders during the problem formulation phase are in danger of not providing answers or insights into issues of importance to stakeholder groups. For example, a key stakeholder such as the Izaak Walton League (http://www.iwla.org/) may view potential impacts to fish-eating birds to be the area of greatest concern, yet the risk assessment may focus exclusively on songbird impacts. The problem formulation would provide reasoning as to why the analysis focused on songbirds and not fish-eating birds. Communication occurs at various stages throughout the risk assessment and prior to further iterations of the risk assessment (Table 8.2). The risk management step will integrate other information relevant to decision making (e.g., legal mandates; political, social, and economic considerations; benefits analyses).

Stakeholders must be identified and engaged in the risk assessment during the problem formulation stage. Improving the flow of information and increasing opportunity for critical discussion among stakeholders involved in the risk communication process will improve the quality of the risk assessment and risk management decisions (Warren-Hicks and Moore 1998). Early interaction during the problem formulation with stakeholders will also increase trust and credibility in the process.

Table 8.2 Process for communication at various stages of the risk assessment

Step in process	Communication activities
1 Problem formulation	Dialogue between risk assessor, decision maker, and stakeholders
2 Analysis	Dialogue between risk assessor, decision maker, and other experts (e.g., academics) as necessary to complete the analyses
3 Risk characterization	Risk assessor communicates results to decision maker
4 Further iterations of the ecological risk assessment	Risk assessor communicates results to decision maker
5 Risk management	Communication from risk assessor and decision makers to stakeholders
6 Implementation of risk management	Dialogue between decision maker and stakeholders

Complex risk assessments, such as those required in situations where benefits and/ or potential risks are high and/or uncertainty is great, will require greater and better informed interaction than less complex risk assessments where risk outputs fall clearly within the categories of acceptable or unacceptable risk. In the complex case, a series of meetings may be required to ensure adequate interaction and involvement of stakeholders. When a decision is made to conduct a probabilistic risk assessment, it is important to help stakeholders understand the principles of probabilistic risk assessment (USEPA 1999) and the rationale for taking this approach. Providing ongoing opportunities for stakeholders to become familiar with the probabilistic risk assessment process will allow increased participation and broaden opportunities for shared, constructive debate.

Decision makers and others responsible for communication to stakeholders should attempt to identify the risk perceptions, concerns, level of probabilistic risk assessment understanding, and information requirements for each sector of stakeholders before and during the problem formulation stage. This knowledge can help identify the risk and uncertainty information to be communicated and the most effective method of communication. The stage of the risk assessment and risk management process at which each stakeholder sector can meaningfully contribute should be made clear, and frameworks and fora established to ensure exchange of information and dialogue.

The key information that should be communicated to each stakeholder sector needs to be identified together with the method of communication. Stakeholder knowledge and ability to assign resources for developing familiarity with probabilistic risk assessment will vary greatly. Also, the interests of stakeholders may be focused on specific sets of information (e.g., risk to a particular resource or species). The tailoring of information to specific needs and level of expertise is critical to ensuring that effective risk communication takes place at each level of participation. At the same time, information should be easily accessible for participants who wish to pursue greater levels of precision and complexity. A well thought-out communication plan is essential to an effective and inclusive process. Warren-Hicks and Moore (1998) detail some of the general rules and steps to successful communication of risk and uncertainty.

8.5   RISK ASSESSOR AND DECISION MAKER ROLES AND RESPONSIBILITIES

Communicating in a manner that accurately portrays risk and the nature of confidence in results is integral to and a major challenge for practitioners of probabilistic risk assessment. Inadequate communication of scientific uncertainty about the effect, severity, or prevalence of a hazard tends to increase unease among decision makers, stakeholders, and other participants. Efforts of the risk assessors should provide clarity for decision makers who must in turn bear ultimate responsibility for communicating the parameters of any decision. The risk assessor will work within a framework that must be clearly communicated by decision makers during the problem formulation. Decision makers at the outset of the risk assessment process must articulate the following points:

Risk assessors will work within an established framework that will include the following:

- Outlining protection goals including resources to be protected and assessment endpoints
- Instituting a regulatory decision-making process including science policy relating to regulatory levels of concern and triggers for further levels of refinement
- Defining the characteristics of each level of refinement of the risk assessment and identifying which levels are deterministic and which are probabilistic
- Identifying the assumptions and uncertainty factors or safety factors at each level of refinement of the risk assessment
- Determining whether the objective of the risk assessment is to provide the most accurate assessment of risk or a protective assessment resulting from the inclusion of conservatively biased uncertainty factors

A risk assessor's job has many dimensions. The 1st is to help the decision maker formulate the core issues or problems in a way that best informs the decision. The decision maker needs help in translating generalized concerns about ecological impacts into specific assessment endpoints for the assessment. Risk assessors need to fully understand the questions being asked by the decision maker to ensure that the risk assessment analysis plan will deliver results necessary for effective decision making. Second, the risk assessor must use scientifically informed technical expertise and judgment to evaluate data and select risk assessment methods and tools appropriate to the problem and available data. Finally, risk assessors are responsible for communicating results, methods, and judgments made throughout the assessment process to the decision maker. They also must communicate to the decision maker 1) risk in relation to assessment endpoints, 2) what effects might occur and their likelihood and magnitude, 3) temporal and spatial occurrence of effects, and 4) confidence in the risk assessment (uncertainty) including data gaps. It is the responsibility of the risk assessor to ensure that uncertainty and risk are explicit and accurately portrayed in the final risk characterization. The decision maker's ability to make decisions and communicate them, in turn, effectively to stakeholders and participants is wholly dependent on interaction and exchange information with the risk assessor.

Decision makers must develop regulatory options and the consequences of these options, taking into consideration public health and ecological, economic, social, and political values. They must manage the complexity of the assessment and the demands of stakeholders within their own regulatory framework and legal mandate. They must be able to communicate the sources and causes of the risk (e.g., nature and intensity, spatial and temporal scale, and recovery potential), their own degree of confidence in the risk assessment, the rationale for risk management decisions, and options for reducing risks (USEPA 1995).

Risk assessors and decision makers both need to be prepared to communicate risk results in an understandable form to other practitioners (regulatory and registrant), stakeholders, and the public. This is particularly critical in the case of uncertainty in the assessment. Most scientists hired to perform risk assessment are thoroughly trained in their subject matter but less familiar with the demands of public presentation or the essentials of educating at multiple levels. Regulators must provide scientists and decision makers with the support and opportunity to develop skills necessary to effectively communicate with stakeholders and the public.

8.6   COMMUNICATION OF UNCERTAINTY FOR REGULATORY DECISION MAKING

Uncertainty occurs under various guises at every stage of the risk assessment and must be integrated during the final risk characterization. Risk practitioners responsible for formalizing the assessment and the resulting decisions must ensure the assessment process is transparent and that risk outputs and uncertainty are effectively communicated. Appropriately applied, uncertainty analysis describes the degree of confidence in the assessment (USEPA 1998) and enables the decision maker to focus risk management decisions.

Irrespective of the risk, assumptions and decisions will have to be made because of uncertainty. Implications of attempting to characterize all variability and uncertainty in the risk assessment need to be considered. For example, exaggerating uncertainties can obscure the scientific basis of risk management decisions, leaving the impression that the decision has been arbitrary in nature (NRC 1989). The purpose of the uncertainty factor together with the type of assessment (e.g., deterministic or probabilistic, protective or best estimate) must be clearly communicated. Uncertainty factors can be described in 3 categories:

1) An uncertainty factor that leads to the best estimate of a variable in an assessment that aims to derive the most realistic estimate of risk
2) A conservatively biased uncertainty factor (i.e., a safety factor)
3) An uncertainty factor that may lead to an underestimate of risk

It has been argued that the use of uncertainty factors is equivalent to having decision making or risk management operating within the risk assessment. Others believe that the proper place for safety factors is in the decision-making process because the objective of the risk assessment is to be credible (believable), not pessimistic or optimistic. Because the overall effect of integrating safety in the assessment is difficult to quantify, a practical approach for risk assessments that informs regulatory decision making would be to establish a reasonably conservative problem formulation but then use best estimates for variables throughout the assessment. Another practical approach would be to use safety factors at lower levels of refinement where the purpose is to screen out "very safe" or "very unsafe" chemicals. Alternatively, this screening level assessment may bracket the true risk using uncertainty factors biased in 1 direction for the 1st risk assessment and then biased in the other direction for a 2nd comparative risk estimate. This may expedite the decision-making process and make the overall risk assessment process more efficient by focusing refined risk assessments that require greater resources on chemicals where the risk is threshold, i.e., not clearly acceptable or unacceptable (Figure 8.1). At higher levels of refinement, where a realistic estimate of risk is necessary, there would be a conscious decision to eliminate safety factors from the assessment.

The decision maker needs to determine whether risks are sufficiently well defined (and data gaps small enough) to support a risk management decision. Also, he or she should determine whether the uncertainty is characterized to the extent acceptable for decision making. In other words, the risk management decision is influenced by uncertainty. Decisions made based on outputs from the risk assessment could fall into 3 alternative categories as shown in Figure 8.1. It is important that uncertainty is clearly communicated because it may be unacceptably high (wide) for effective decision making. In this situation, the risk assessment needs to be refined to reduce uncertainty unless the risk curve and its uncertainty bounds distinctly fall within the category of "acceptable risk" or "unacceptable risk."

Practitioners of ecological risk assessments will frequently experience large uncertainty bounds on the estimates of risk. Unfortunately, characterizing and/or reducing uncertainty can be very costly. However, these costs must be balanced with the need to conduct sufficient analysis to make an informed decision.

Selection of the uncertainty analysis method to use in the risk assessment is affected by the utility of the method for risk management purposes, and different approaches may be necessary for different questions posed by the decision maker. The decision maker may require a certain method of uncertainty analysis depending on where the risk falls relative to the threshold of acceptability or unacceptability. Other factors influencing choice of method include the type of regulatory decision (registration versus reregistration) to be made, the stage in the risk assessment that the decision is being made, and consequences of implementing the decision. The method selected must provide the regulator with the necessary information for risk management and also for communicating risk to stakeholders. The risk assessor must be able to justify use of a particular uncertainty analysis method to support the needs of the decision maker.

The information communicated by the risk assessor needs to provide an assessment of the overall degree of uncertainty and confidence in the analysis. The nature of the uncertainty for sensitive variables should be communicated, for example, variability, descriptive errors, data gaps, uncertainty about a quantity's true value, and

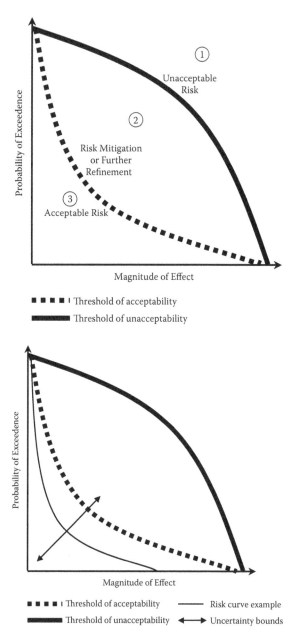

FIGURE 8.1 The risk curve lines shown represent thresholds between different types of decisions (based on ECOFRAM 1999a and 1999b). These thresholds would be determined by decision makers and may move location subject to other factors that affect the decision (e.g., pesticide benefits). The bottom graph shows an example risk curve with uncertainty bounds. The curve clearly fits within the "acceptable risk" category; however the upper uncertainty bound does not, indicating a need for risk mitigation or further refinement of the risk assessment.

uncertainty about model structure and detail. The decision maker must be provided with adequate relevant information that can be used for the purpose of decision making. The statements made concerning risk need to be clear and accurately reflect the data inputs, input distributions, assumptions, and uncertainties. Accordingly, the risk assessors need to stay within the limitations of the supporting data and give due consideration to these limitations as articulated by uncertainty analyses through each step of the assessment.

8.7   REFERENCES

[ECOFRAM] Ecological Committee of FIFRA Risk Assessment Methods. 1999a. US Environmental Protection Agency. Office of .Pesticide Programs, Washington, DC.

[ECOFRAM Ecological Committee of FIFRA Risk Assessment Methods. 1999b. US Environmental Protection Agency. Office of .Pesticide Programs, Washington, DC.

[NRC] National Research Council, Committee on Risk Perception and Communication. 1989. Improving risk communication. Washington (DC): National Academy.

Peterson RKD. 2000. Public perceptions of agricultural biotechnology and pesticides: recent understandings and implications for risk communication. Am Entomol 46:8–16.

Powell D, Leiss W. 1997. Mad cows and mother's milk. The perils of poor risk communication. Montreal (CA): McGill-Queen's University Press.

[USEPA] US Environmental Protection Agency. 1995. Ecological risk: a primer for risk managers. Washington (DC): EPA/734/R-95/001.

[USEPA] US Environmental Protection Agency. 1998. Guidelines for ecological risk assessment. Washington (DC): EPA/630/R-95/002F. Risk Assessment Forum.

[USEPA] US Environmental Protection Agency. 1999. Risk assessment guidance for superfund: volume 3, part A, process for conducting probabilistic risk assessment. Washington (DC): EPA 000-0-99-000. Office of Emergency and Remedial Response.

Warren-Hicks WJ, Moore DRJ. 1998. Uncertainty analysis in ecological risk assessment. In: Proceedings from the Pellston Workshop on Uncertainty Analysis in Ecological Risk Assessment. 23 to 28 August 1995. Pellston (MI). Pensacola (FL): SETAC.

9 How to Detect and Avoid Pitfalls, Traps, and Swindles

G. Joermann, T. W. La Point, L. A. Burns,
J. P. Carbone, P. D. Delorme, S. Ferson,
D. R. J. Moore, and T. P. Traas

9.1 INTRODUCTION

A probabilistic risk assessment (PRA) deals with many types of uncertainties. In addition to the uncertainties associated with the model itself and model input, there is also the meta-uncertainty about whether the entire PRA process has been performed properly. Employment of sophisticated mathematical and statistical methods may easily convey the false impression of accuracy, especially when numerical results are presented with a high number of significant figures. But those who produce PRAs, and those who evaluate them, should exert caution: there are many possible pitfalls, traps, and potential swindles that can arise. Because of the potential for generating seemingly correct results that are far from the intended model of reality, it is imperative that the PRA practitioner carefully evaluates not only model input data but also the assumptions used in the PRA, the model itself, and the calculations inherent within the model. This chapter presents information on performing PRA in a manner that will minimize the introduction of errors associated with the PRA process.

Burmaster and Anderson (1994) have compiled a list of principles of good practice, which were originally aimed at Monte Carlo simulations, but are valuable also for other techniques in uncertainty analysis. These recommendations later appeared in modified and supplemented form in various handbooks and other publications, e.g., the USEPA Guiding Principles for Monte Carlo Analysis (USEPA 1997).

This chapter summarizes the major preconditions necessary for the conduct of an environmentally relevant PRA, illustrates potential sources of errors, and provides recommendations about how to avoid them. The chapter draws upon the Burmaster and Anderson (1994) principles that still form an excellent basis with regard to good practice. Other sources are Ferson (1996), Warren-Hicks and Moore (1998), and Cullen and Frey (1999). Apart from those approaches noted, there are many cross references to other chapters within this volume.

The use of probabilistic techniques is rather new in ecotoxicology, and many issues regarding the appropriate methodology necessary to conduct a PRA are not yet settled. Because of the fluid nature of the process, different options regarding assumptions, procedures, and default values exist. Because of the developmental stage of PRA, there likely will not be a formally standardized approach or generally accepted method in the scientific or regulatory community from which guidance can be sought. It is therefore imperative to exert well-founded expert judgment when making decisions about PRA parameterizations. Within that context, transparency is of pivotal importance. It is always helpful to make explicit all data, models, tools, and procedures. Where appropriate, give references, sources, and documentation, which allows for an assessment of the quality of the PRA in general. Poor input data should be noted, and weak points in models and methods should be fully elucidated.

9.2   MEANINGFUL PROBLEM FORMULATION

Essential prerequisites for a probabilistic risk assessment are a well thought-out problem formulation and a clear definition of the assessment endpoints. The probabilistic approach according to its very nature aims at making predictions on quantities or the occurrence of certain events. Such quantities and events must be specified precisely such that, at least in principle, there is no doubt on what the quantity is or whether the event happened (Morgan and Henrion 1990).

9.3   SUITABILITY OF INPUT DATA

Input data should be described in sufficient detail. That alone, of course, doesn't guarantee their quality but it allows for a judgment whether the data are suitable for the intended purpose. Input data could be weak for a variety of reasons:

- Lack of accuracy
  Accuracy is the distance of a measured or estimated value to its true value. Inaccuracies (systematic deviations or bias) may arise from inadequate survey design and processing of the samples. Expert judgment is needed to assess the accuracy; however, that implies that the origin and the extraction of the data is described (sampling design and conditions).
- Data not representative
  Data must be representative of the scenario to be assessed. If the population and conditions under assessment are different from the data source, or the scope is broader than the data source, then caution must be exercised. In those cases it may be possible to demonstrate representativeness by bridging data or other supporting evidence. With regard to effects data, extrapolation from lab to field is often involved, thus introducing an additional degree of uncertainty.
- Data range exceeded
  Where regression statistics are involved the relationship should not be extrapolated beyond the observed data range for the independent variable.

This is particularly important if the linear regression model is used rather than a mechanistic model.
- Small sample size
  With small sample sizes the uncertainty due to random sampling error usually is large and may become the dominant source of uncertainty in the output. This uncertainty could be reduced if there is relevant prior information, for example, reasonable estimates for distribution parameters from well-described datasets (Aldenberg and Luttik 2002).

9.4 PARAMETERIZATION OF THE DISTRIBUTION OF INPUT VARIABLES

The selection of the appropriate distribution type for a model variable is of critical importance. Selection of spurious input parameter distributions while allowing the generation of sophisticated "looking" output distributions would skew the model outcome away from accuracy. In order to demonstrate proper parameterization of the input distribution, the model practitioner is urged to plot the data points versus the distribution function. Goodness-of-fit statistics should be reported, but it must be kept in mind that the power of such tests depends upon the sample size. For small sample sizes the null hypothesis often cannot be rejected although the fit is poor, and with large sample sizes the null hypothesis may be rejected although the deviation from the fitted curve is unimportant from a practical perspective. Therefore a visual check of the graph image is always helpful. There are no simple rules on how to choose from different distribution types; however, goodness-of-fit tests should not form the sole basis for decisions regarding the appropriate selection of distribution shape. Instead, the practitioner should consider whether PRAs have been conducted with similar kinds of data or whether underlying mechanisms are known that suggest a certain type of distribution.

- Incorrect choice of distribution
  It is key to select distributions that appropriately represent the data. As an example, one should not use the Poisson distribution to describe continuous data. Use of a uniform distribution when information is scant is typical; however, the approach is potentially a pitfall because it assumes a high degree of certainty about the distribution minimum and maximum values when in reality those values may not be certain.
- Constraints on variables not observed
  Some variables cannot be negative (concentration, body weight); other variables have upper bounds (e.g., 100%). If the fitted distribution exceeds these bounds the tails may be truncated (draws in a Monte Carlo analysis have to be processed accordingly); however, distributions that have to be severely truncated are a poor choice. Especially proportions or fractions that range between 0 and 1 (0% and 100%) should only be represented by a distribution with finite tails (e.g., beta or uniform distribution).

9.5 CORRELATIONS AND DEPENDENCIES

In Monte Carlo and other techniques, correlations between input data have an influence on the shape of the output distribution, especially on the tails (positive correlations usually will cause wider tails, negative correlations narrower tails). The tails of the output distribution are generally important with regard to the ultimate conclusions drawn from the analysis. Incorrect accounting for correlations can be a serious source of misleading results. If submodels are fitted to data (i.e., linear or nonlinear regression), the parameters are necessarily correlated. A good statistics program gives the covariance matrix (from which correlations can be calculated) or the correlations between the parameters of the fitted model.

- Problem of dependencies ignored
  Input parameter dependencies are frequently ignored because typically there is a lack of information regarding dependencies. However, it is an essential requirement that input parameters are appropriately addressed in the PRA. If specific data are not available, the PRA practitioner must employ expert knowledge to judge whether a certain degree of correlation is plausible. Because the impact of input parameter dependencies on the outcome of the PRA is unknown, sensitivity analyses should be performed where simulations are run assuming no, moderate, and high correlations. Approaching the matter in this fashion will ensure that time and resources are used appropriately.
- Known dependencies not considered
  Information regarding model input correlations must be accounted for if data or sensitivity analyses indicate that correlations are highly influential on model outcomes. In cases where there exists a strong dependence between variables, the model could be modified by holding 1 variable fixed while taking the 2nd as a random variable. For example, the fraction of a pesticide reaching the soil (f_s) is obviously inversely correlated with the fraction intercepted by the vegetation (f_i). Using that assumption, f_i could be regarded as a random variable and f_s defined as $1 - f_i$. There are additional approaches that can be employed to cope with dependencies either in the model or in the method for sampling in a Monte Carlo analysis.
- Correlation matrix nonsensical
  If more than 2 variables are involved, the correlation matrix must be positive, semidefinite. That means if the correlation with A and B is a, and the correlation between B and C is b, then the correlation between A and C can't be any number between 0 and 1; it must satisfy certain constraints. Such errors may occur if pairwise correlation coefficients stem from different data sets.

9.6 MODEL UNCERTAINTIES

The goal of any model simulation is to reasonably represent a snapshot of reality. With quantitative exposure or risk assessment there is typically a range of input variables that are linked to give 1 or more output variables. The system may contain submodels where the output of one is fed to the input of another.

Validation of models is desired but can be difficult to achieve. Models are empirically validated by examining how output data (predictions) compare with observed data; (such comparisons, of course, must be conducted on data sets that have not been used to create or specify the model). However, model validations conducted in this manner are difficult given limitations on data sources. As an alternative approach, model credibility can be assessed by a careful examination of the subcomponents of the model and inputs. One should ask the question: Does the selection of input variables and the way they are processed make sense? Also, confidence in the model may be augmented by peer reviews and the opinion of the scientific community. Common faults and shortcomings are

- Inappropriate choice of model
  Usually models are created for a certain purpose, and that purpose drives their structure, level of detail, level of complexity, etc. A model may be excellent, but it must not be used for inappropriate purposes. If the output of the model does not match the assessment endpoint and the questions raised in the problem formulation phase, then the model obviously is not suitable for the specific case.
- Overparameterization of empirical models
  As a general rule, the number of parameters in data-based models should be kept at a minimum. Increased input parameterization may improve model accuracy, but at a cost, since it requires greater time and resource commitments. Increasingly, sophisticated models require greater precision with regard to input parameterization. Without the greater attention to input veracity, the modeling practitioner runs the very real risk of generating increasingly meaningless output. Additional parameters improve the fit, but that is irrelevant. Only when there is a significantly better fit are more parameters justified.
- Model boundaries exceeded
  The validity of a model is always limited to a certain domain in the parameter space. For example, if a quantitative structure-activity relationships (QSAR) model is specified for nonpolar organic chemicals in the log P_{ow} range from 2 to 6 and has a molecular weight below 700, then an application to substances outside this range is an improper extrapolation. Note that the parameter space may be difficult to discern; for example, combinations of low values for one variable and high values for another could constitute an extrapolation if such combinations had been missing in the validation or specification of the model. Exceedence of model boundaries introduces additional uncertainty at best, but can also lead to completely incorrect outcomes.
- Hidden assumptions
  If a model is based on incorrect assumptions then the output cannot reflect reality. Therefore, it is essential to make explicit all assumptions and settings used in defining the scenarios, e.g., spatial and temporal dimensions

of the system, number of individuals in the system, etc. Such parameters should appear as design variables with fixed values in the list of input variables. It should not be assumed that design variables are without uncertainty; usually they are deliberately set to fixed values in order to keep models simple.

- Biological or ecochemical constraints ignored
  In addition to trivial constraints on single variables (survival rate must be between 0% and 100%) general biological knowledge may dictate more rules for data sets, e.g., in population models the number of emigrants cannot be higher than the total population. There also may be cases where certain combinations of 2 inputs are nonsensical; if that is true such combinations should be excluded in simulations, for example, by establishing families of data sets (binning).
- Dimensions and units not concordant
  Mathematical equations must balance dimensionally and permit concordance among the units involved.
- Data lumped
  If input data are differentiated for subpopulations, the between-group variance should be examined before pooling the data.
- Spatial and temporal resolution not appropriate
  Input variables usually have a structured pattern in space and time. It can be described as autocorrelation (mostly variates tend to be more similar to their neighbors in space and time than to distant observations) or as a partitioning of the variance into hierarchical levels (sample–plot–field, or hour–day–month). The lower limit of resolution always is dictated by the data sampling protocol. In a model, a fine resolution may be made coarse by averaging. The reverse is not true, i.e., coarse resolution can not be refined by dividing. In exposure models the temporal resolution is to be seen in relation with the response characteristic of the effect endpoint under consideration (e.g., a 1-week-average concentration in surface water is inappropriate for combining with a fast lethal effect). The spatial resolution is to be seen in relation with the home range or foraging range of the individuals or populations under consideration. Generally, variability should be processed in such a way that it matches the unit of analysis.

9.7   SOFTWARE TOOLS AND COMPUTATIONAL ISSUES

PRA models are usually implemented in a computer program, which can be a simple spreadsheet or more complex models in specific programming languages. This process may lead to errors. There simply is the possibility that the computer program does not perform as it was meant to.

- Computerized version of model faulty
  It must be ensured that the conceptual model is correctly translated into the mathematical notation, and that in turn into the computer code. Efforts to check for mistakes (test runs of example data sets, inspection of intermediate

results, and cross checks with other software) should be increased proportionately to the complexity of the newly developed computer code.

- Random number generator flawed
  In modern software packages random number generators should work satisfactorily. However, if seeds are set manually, procedures should ensure that repetition of sequences is avoided.
- Multiple instantiation of the same variable in a Monte Carlo analysis
  Sometimes a variable appears more than once in a model (e.g., in submodels or in additive terms). For all computational steps addressing the same unit of analysis (e.g., exposure of an individual bird) a single instantiation of the variable should be used, that means a variate must be drawn only once in each replicate of the Monte Carlo simulation.
- Sampling method in Monte Carlo analysis inappropriate
  There are several sampling techniques in Monte Carlo analyses, the most common being random, median Latin hypercube and random Latin hypercube. Latin hypercube techniques are usually preferred because they need fewer iterations and thus are more efficient. They are, however, inferior to random sampling if high percentiles of the output are of interest and if the exact shape of the output distribution is important (Cullen and Frey 1999).
- Number of iterations in Monte Carlo analysis too low
  The number of sampling iterations must be sufficient to give stable results for output distributions, especially for the tails. There are no simple rules, because the necessary number of runs depends on the number of variables entered as distributions, model complexity (mathematical structure), sampling technique (random or Latin hypercube), and the percentile of interest in the output distribution. There are formal methods to establish the number of iterations (Cullen and Frey 1999); however, the simulation iterations could simply be increased to a reasonable point of convergence.
- Incorrect scale conversions
  If, for example, an exposure estimate is scaled down by a factor of 2 (because the application rate per ha is halved), then it is correct to divide arithmetic means and standard deviation by 2, but it is not correct to divide logarithms of mean and standard deviation by 2.

9.8   PRESENTATION AND INTERPRETATION OF RESULTS

The output of a PRA is always conditional with regard to the input data. That is the reason why it is so important to mention all data assumptions and not to withhold limitations and information gaps. The description of the results and the interpretation should be kept apart. The former is the faithful translation of the mathematical outcome into plain language, the latter is a discussion on what the result means and what conclusions can be drawn.

- Uncertainty and variability confused
  Uncertainty and variability should be treated as distinct entities. Both can be handled in 1 assessment, and can be represented as the result of a

2-dimensional simulation in 1 graph, but interpretations should be clear-cut.

- Missing details
  Available information regarding subpopulation probabilities should be always included. As an example, a certain effect evident on aquatic organisms in 2% of all water bodies may or may not be significant. However, indicating that that same effect is apparent in 80% of water bodies of a certain type may influence the significance of that data.
- Verbal description of the results unprecise
  Results should be described precisely. It is important for which entity a certain probability holds. Assume surface water intake is represented by a distribution. The mere information that there is a 2% probability that a predefined ecologically acceptable concentration is exceeded is too scanty. Rather, it must be stated whether 2% of all water bodies in a certain area receive that intake, or 2% of the water bodies adjacent to any agricultural fields, or 2% of the water bodies adjacent to treated fields, or whether 2% of treatments will result in a contamination of any water body. Furthermore, results should not be overly condensed.

9.9   CONCLUSIONS

A probabilistic risk assessment is a complex undertaking that typically involves a multitude of input data, assumptions, models, and computational tools. That is why the execution of such analyses is error prone. In order to avoid mistakes and pitfalls successfully, assessors should first provide a detailed documentation of data, models, and procedures. Although this will not ensure reliability, it helps both creators and reviewers of a PRA to identify potential shortcomings. While it always should be possible to remove technical faults, there might be issues where it is difficult or expensive to find a remedy because information, e.g., on dependencies, simply is lacking or because input data and assumptions, although recognized as crude already, are the best available. The minimum requirement in such cases is to analyze to what degree the result of a PRA is influenced by the parameter in question, and if it turns out to be an uncertainty then carry on this information up to the result.

9.10   REFERENCES

Aldenberg T, Luttik R. 2002. Extrapolation factors for tiny toxicity data sets from species sensitivity distributions with known standard deviation. In: Posthuma L, Suter II GW, Traas TP, editors. Species sensitivity distributions in ecotoxicology. Boca Raton (FL): Lewis Publishers, CRC Press.
Burmaster DE, Anderson PD. 1994. Principles of good practice of the use of Monte Carlo techniques in human health and ecological risk assessment. Risk Anal 14:477–481.
Cullen AC, Frey HC. 1999. Probabilistic techniques in exposure assessment. New York and London: Plenum Press.
Ferson S. 1996. Automated quality assurance checks on model structure in ecological risk assessment. Human Environ Risk Assess 2:558–569.

Morgan MG, Henrion M. 1990. Uncertainty: a guide to dealing with uncertainties in quantitative risk and policy analysis. Cambridge (UK): Cambridge University Press.

[USEPA] US Environmental Protection Agency. 1997. Guiding principles for Monte Carlo analysis. Risk Assessment Forum. Washington (DC): USEPA, Document EPA/630/R-97/001. Available from: http://www.epa.gov/ncea/monteabs.htm

Warren-Hicks WJ, Moore DRJ, editors. 1998. Uncertainty analysis in ecological risk assessment. Society of Toxicology and Chemistry (SETAC) Pellston Workshop on Uncertainty Analysis in Ecological Risk Assessment; 1995 Aug 23 to 28; Pellston (MI). Pensacola (FL): SETAC.

10 Conclusions

A. Hart, T. Barry, D. L. Fischer, J. M. Giddings,
P. Hendley, G. Joermann, R. Luttik, D. R. J. Moore,
M. C. Newman, E. Odenkirchen, and J. L. Shaw

10.1 INTRODUCTION

In the Introduction to this book, we listed several key issues addressed at the workshop in Pensacola. This chapter presents our conclusions.

In some cases, the workshop produced a near consensus on how to resolve the issue, but further work is required to confirm and implement the conclusion. In most cases, however, the workshop has identified a range of possible solutions and further work is required to evaluate them. Uncertainty analysis in pesticide risk assessment is highly encouraged; however, uncertainty analysis should be used and interpreted with caution. The methods used should be justified and described in detail in every assessment.

10.2 WHICH METHODS OF UNCERTAINTY ANALYSIS ARE APPROPRIATE UNDER WHAT CIRCUMSTANCES?

The workshop reviewed 7 contrasting methods of analyzing uncertainty in risk assessments:

- Bayesian inference
- First-order error analysis
- First-order (nonhierarchical) Monte Carlo
- Second-order (hierarchical or 2D) Monte Carlo
- Bayesian Monte Carlo
- Interval analysis
- Probability bounds analysis

Other methods exist and may deserve more consideration.

The workshop did not reach firm conclusions on which methods of uncertainty analysis are suitable for use in pesticide risk assessment, or when they should be used.

Experts in uncertainty analysis hold differing opinions on the merits of the various methods, partly as a result of differing theoretical perspectives. These differences are likely to continue and may be healthy from a scientific standpoint, e.g., as a stimulus to further advances.

Nevertheless, from the standpoint of practical regulatory assessment, it would be desirable to reach a consensus on the selection of methods for routine use. Although there are practical benefits to agreeing on a limited number of methods for routine use, there may be scientific reasons for preferring alternative methods in particular cases. In order to avoid unusual demands on regulators to absorb a wide array of uncertainty methods, novel approaches should only be presented alongside more familiar (conventional) approaches with an explanation of how the new approach improves the risk assessment. In addition, it should be recognized that novel approaches are likely to take longer for regulatory review.

10.3   WHAT ARE THE IMPLICATIONS OF PROBABILISTIC METHODS FOR PROBLEM FORMULATION?

Formulating the assessment problem well is an essential foundation for risk assessment. The workshop considered how the use of probabilistic models and uncertainty analysis affects problem formulation and its main components: the integration of available information, definition of the assessment endpoint, specification of the conceptual model, and planning of the analysis phase.

The workshop concluded that the use of probabilistic methods requires increased attention to the following aspects of problem formulation:

1) Define the assessment endpoint precisely, in terms of probabilities, e.g., the probability of a given level of mortality in the exposed population.
2) Ensure that the assessment endpoint is capable of being modeled and has attributes that are measurable.
3) Ideally, define the assessment endpoint so that it relates directly to the management goal. If this is not practical, (e.g., if the management goal refers to population sustainability but the assessment endpoint refers to individual mortality), define in advance how the assessment endpoint will be interpreted. If this involves subjective judgments then consider the use of formal methods.
4) Explicitly define the mechanisms and spatial, temporal, and biological dimensions of the system assessed to an appropriate, but not excessive, level of detail. Beware of inappropriate aggregation that may distort or hide important effects.
5) Systematically identify, evaluate, and incorporate the major sources of uncertainty, including model uncertainty. Initially, all potentially significant routes of exposure and types of effect should be included. Identify models to represent these processes. Use sensitivity analysis to identify insignificant variables, exposure routes, and effects.
6) Take advantage of opportunities to reduce the variability in individual assessments by defining separate scenarios. Make the specification of scenarios a distinct step in problem formulation.
7) Choose appropriate methods of uncertainty analysis and consider their implications for other aspects of problem formulation.
8) Be more effective in gathering and incorporating other lines of evidence.

9) Plan at the outset how one will communicate the results, both to decision makers and other stakeholders.

10) Have a clear vision of the roles of the risk assessor and decision maker and ensure they interact efficiently throughout the process.

11) Take advantage of opportunities to devise generic problem formulations, but check their appropriateness for each individual assessment.

12) Consider with stakeholders the uncertainties in risks, costs and benefits, and the consequences of false positives and false negatives when establishing decision rules.

10.4   HOW CAN UNCERTAINTY ANALYSIS METHODS BE USED EFFICIENTLY AND EFFECTIVELY IN DECISION MAKING?

Reducing uncertainty is usually expensive. Therefore, an iterative process should be used to conduct the minimal amount of analyses that are necessary to characterize and reduce uncertainty to the point where an informed decision can be made. The point where a decision can be made with acceptable uncertainty will depend on the "threshold of acceptability" and "threshold of unacceptability."

The type of decision that needs to be made will influence the choice of uncertainty analysis method. Consequently, the process must include a dialogue between the risk assessor and decision maker throughout the risk assessment. The uncertainty associated with the risk assessment must be clearly communicated so that all parties involved in the risk assessment process understand it.

In a screening-level risk assessment, interval or bounding analyses, which put upper and lower bounds on risk, may be sufficient to reach a decision of "acceptable risk" or "unacceptable risk" provided the bounds are a reflection of the true limits of uncertainty.

A process is outlined for reaching the desired level of certainty while minimizing resource requirements and maximizing efficiency:

1) A carefully planned problem formulation needs to be developed and implemented. The analysis plan of the problem formulation will outline the uncertainty analysis methods to be used.

2) Use sensitivity analysis to determine which parameters are the driver of the model (e.g., Dakins et al. 1994).

3) Determine where additional data will reduce uncertainty the most. The process of Monte Carlo and 1st-order error analysis can help with sensitivity analysis and help to identify variables that need refinement and better data.

4) Determine whether there are more cost-effective alternatives to additional data generation and risk assessment refinements. What-if analyses can be used to examine the savings in risk management that might result from additional data generation. Techniques that may be suitable for this include Bayesian Monte Carlo and expected value of information (EVOI) analysis (Dakins et al. 1996).

5) Conduct risk assessment refinements using appropriate uncertainty analysis methods based on output from the sensitivity analysis. Use appropriate experts in this process.

10.5   WHEN AND HOW SHOULD WE SEPARATE VARIABILITY AND UNCERTAINTY?

There is variability in exposure and effects of pesticides in the real world: they vary over space and time, and between biological entities (between individuals, between species, etc). There is uncertainty* in our knowledge of exposure and effects and of the parameters used to model them.

Some approaches to uncertainty analysis (e.g., 2D Monte Carlo and P-bounds) enable the assessor to separate variability and uncertainty. Other approaches do not separate them, and some schools of thought regard the distinction between variability and uncertainty as artificial or unhelpful.

Risk managers are interested in both variability and uncertainty: they want to know how the expected impacts will vary (how frequent and widespread will impacts be?), and they want to know how certain the assessment is (how sure are you, what are the confidence limits?).

Risk managers may need assessors to separate variability and uncertainty explicitly, if they have different implications for decision making. For example, 100% certainty that 10% of individuals will die is likely to have different implications from a 10% chance that 100% of individuals will die.

Furthermore, separating variability and uncertainty can help risk managers and assessors to decide whether to collect additional information and, if so, on which parameters. This is because uncertainty can be reduced by obtaining additional information, but variability cannot. If there is little uncertainty, then the effects are already well characterized and obtaining further data will make little difference to the assessment outcome. If there is much uncertainty, then priority should be given to obtaining better information about those parameters from which it mostly derives.

Therefore, from a practical regulatory viewpoint, there are substantial advantages in separating variability and uncertainty. These advantages apply generally, with 1 exception. If a screening assessment shows that the likelihood of effects is acceptably low even when variability and uncertainty are combined, then there is no benefit in separating them because the interpretation is clear already, and no further data collection is required. In all other assessments, separation is potentially helpful.

In assessments where variability and uncertainty are separated, it may not be necessary to separate them for every input parameter. If sensitivity analysis shows a parameter has little influence on the assessment output, then variability and uncertainty for that parameter need not be separated (indeed, it could be treated as a constant).

---

* Some authors use "incertitude" for limitations on knowledge, and "uncertainty" as a collective term that includes both variability and incertitude. Most of the workshop participants preferred to use the term "uncertainty" in the more specific sense, referring to limitations on knowledge.

Currently, approaches that separate variability and uncertainty have rarely been used for pesticide assessments, so further evaluation is needed to determine whether they are unsuitable for other reasons (e.g., complexity or cost). Also, it can be difficult to separate variability and uncertainty in real datasets, so the development of guidance on this would be helpful.

10.6   HOW CAN WE TAKE ACCOUNT FOR UNCERTAINTY CONCERNING THE STRUCTURE OF THE RISK MODEL FOR THE ASSESSMENT?

Uncertainty about the form of a model results when there is disagreement within the scientific community about the underlying processes, when the underlying mechanisms are poorly characterized, or when extrapolation beyond existing data or theory is necessary.

Several approaches exist for dealing with model uncertainty:

- One-at-a-time (OAT) method or scenario analysis: alternative models are analyzed separately and the results are compared.
- Model weighting: different models are combined by assigning weights representing their relative probability, using either Bayesian and non-Bayesian approaches.
- Meta-models: a global model is developed that contains plausible models as special cases, converting model uncertainty into uncertainty about model parameters. Again, this can be done using either Bayesian and non-Bayesian approaches.
- Model enveloping: the outputs from alternative models can be combined using bounding methods (e.g., probability bounds analysis).

The workshop recognized the importance of dealing with model uncertainty but did not evaluate the alternative approaches in detail. Further work is required to identify instances of model uncertainty for pesticide risk assessment and to develop guidance on how to deal with it.

10.7   HOW SHOULD WE SELECT AND PARAMETERIZE INPUT DISTRIBUTIONS WHEN DATA ARE LIMITED?

Risk assessors often encounter situations in which the available datasets may appear, on 1st consideration, to be of limited capacity to support the parameterization of distributions for a given risk assessment model variable.

An initial step in addressing such situations should be the performance of an analysis of the sensitivity of a risk assessment model to changes in the variable. If the model proves relatively insensitive to conservative bounds to the variable, then further consideration of uncertainty for this variable may be unnecessary and a point estimate may suffice.

If the risk assessment model is found to be sensitive to the variable in question, a number of options are available to address its parameterization and may include the following:

- Refer to well-described distributions from other cases in order to understand default shape characteristics and use this information to develop bounds or parameter estimates for the present case.
- When information is severely limited (e.g., range data, summary statistic, or limited quantiles), 1 option is to apply a maximum entropy approach to distribution parameterization.
- Consider use of Kolmogorov–Smirnov intervals to explicitly calculate uncertainty.
- Consider statistical approaches to estimate variance on the basis of sample size.
- Apply information regarding underlying mechanistic processes associated with the variable (chemical, physical, or biological) that may suggest appropriate distribution families.
- Implement hierarchical approaches, along with professional judgment, and reference to other cases, to account for uncertainty in the estimation of distribution parameters.

10.8   HOW SHOULD WE DEAL WITH DEPENDENCIES, INCLUDING NONLINEAR DEPENDENCIES AND DEPENDENCIES ABOUT WHICH ONLY PARTIAL INFORMATION IS AVAILABLE?

The possibility of dependencies should be considered in every assessment. Where dependencies are highly unlikely, assume independence. If you have enough information, then include the dependencies in the model. Where dependencies are possible, but information to quantify them is limited, conduct what-if analyses to explore their possible consequences.

Possible approaches include

- Rewrite the model, modifying the structure so as to reduce the number of variables that are highly correlated with one another.
- Run the model several times, once assuming independence and again assuming plausible types and degrees of correlation, based on prior knowledge or possible mechanisms. Compare the results.
- Model different scenarios separately, or use families of information ("binning" of inputs, e.g., soil properties) so as to ensure that dependencies are accounted for.
- Use Bayesian or non-Bayesian updating to infer the correlation structure.
- Use methods such as bounding analysis that do not require assumptions about dependencies among inputs.

10.9   HOW CAN WE TAKE ACCOUNT OF UNCERTAINTY WHEN COMBINING DIFFERENT TYPES OF INFORMATION IN AN ASSESSMENT (E.G., QUANTITATIVE DATA AND EXPERT JUDGMENT, LABORATORY DATA, AND FIELD DATA)?

A key issue is how to account for uncertainty when diverse types of evidence are being applied to a single parameter or the entire assessment. Field data and laboratory data are combined in complex ways. For example, predictions based on laboratory-derived 1st-order constants for the degradation of a pesticide might be combined with observations from a field-based time course study of pesticide disappearance from a pond. Quantitative estimates with defined uncertainties might be combined with qualitative insight. As an illustration, Monte Carlo simulation of avian exposure to a pesticide might be conducted with an informed opinion that imbibing pesticide in drinking water may or may not be important.

This issue was generally agreed to be important, and several suggestions were made. An overarching recommendation was that formal methods are preferred to ad hoc procedures. All relevant information should be provided, including graphical representations where possible, in order to maximize understanding during the melding of information.

Several examples of applicable formal methods were discussed. Nonhierarchical quantitative models can be applied several times based on plausible scenarios emerging from a qualitative informed opinion. Expanding on the example above, a Monte Carlo simulation of pesticide ingestion rates may be conducted with and without consideration of water sources. Hierarchical Monte Carlo methods can be used in a similar manner.

Bayesian methods are very amenable to applying diverse types of information. An example provided during the workshop involved Monte Carlo predictions of pesticide disappearance from a water body based on laboratory-derived rate constants. Field data for a particular time after application was used to adjust or update the "priors" of the Monte Carlo simulation results for that day. The field data and laboratory data were included in the analysis to produce a posterior estimate of predicted concentrations through time. Bayesian methods also allow subjective weight of evidence and "objective" evidence to be combined in producing an informed statement of risk.

Regardless of the method used, the basis of the final risk characterization must be explicit. All components and sources of evidence should be described. The explicit linkage between the analysis results and the assessment endpoints must be clearly but adequately stated. Tandem presentation of conventional methods (e.g., ad hoc weight of evidence) and formal methods (e.g., Bayesian, meta-analysis) are recommended to enhance understanding. This is intended to facilitate acceptance of unfamiliar approaches, not to imply that the conventional methods are a touchstone.

10.10   HOW CAN WE DETECT AND AVOID MISLEADING RESULTS?

Uncertainty analyses often employ complex models and sophisticated mathematical and statistical techniques. That is why the execution of such analyses is error prone and the results are susceptible to misinterpretation.

In order to avoid misleading results some key recommendations can be given:

- Clear concept
  Make sure that the problem is thoroughly formulated and the assessment endpoints precisely stated.
- Documentation
  It is essential to make explicit all data, models, tools, and procedures. Where appropriate, references and sources should be given. That alone, of course, does not ensure an appropriate risk estimate (quality of input data satisfactory, models appropriate, etc.), but it helps both creators and reviewers to identify weak points.
- Clear communication of results
  It is essential to fully and precisely describe the characterizations of effects, exposure, and risk to avoid misconceptions about the scope of the final results.
- Other lines of evidence
  Check whether they are consistent with the analysis, and combine them formally where possible.
- Plausibility
  Check whether any assumptions, conclusions, or intermediate outputs conflict with common sense or biological plausibility.
- Apply best practice principles.

10.11   HOW CAN WE COMMUNICATE METHODS AND OUTPUTS EFFECTIVELY TO DECISION MAKERS AND STAKEHOLDERS?

Effective communication among risk assessment practitioners, decision makers, and other stakeholders is essential and is the responsibility of all parties involved in the risk assessment process. Poor communication together with ambiguous, inexplicit, or inaccurate interpretations of risk assessment outputs result in 1) an erosion of scientific credibility, 2) ineffective decision making, and 3) futile use of resources.

Communication between risk managers, risk assessors, and analysts is essential from the start of the assessment process, not just in communicating results. For example, the choice of uncertainty analysis methods will be dependent on 1) the questions posed by decision makers, 2) the closeness of the risk estimate and its bounds to thresholds of acceptability or unacceptability, 3) the type of decision that must be made, and 4) the consequences of the decision.

The workshop favored the use of graphical representations that combine the key elements of the assessment outcome: the magnitude and frequency of effects, together with appropriate confidence bounds. This should always be accompanied by text explaining what the risk estimate and its bounds represent and listing any mechanisms, dependencies, or uncertainties it excludes.

Successful communication will be dependent on the output from the risk assessment being modified and presented in different ways as appropriate for the recipients of the information (i.e., their interests, questions, perspectives, type of involvement in the process, and technical knowledge). The objective is to provide recipients of the message with adequate information to enable them to make their own decision.

Probabilistic techniques are relatively new in the pesticides arena and are prone to criticism and debate. It is therefore especially important to clearly communicate the approach taken, and to explain to decision makers how the risk estimates were quantified.

10.12   WHAT ARE THE PRIORITIES FOR FURTHER DEVELOPMENT, IMPLEMENTATION, AND TRAINING?

1) Further evaluation of alternative methods of uncertainty analysis

For the purposes of practical regulatory assessment, it is desirable to reach a consensus on which methods of uncertainty analysis should be adopted for routine use. This has not yet been achieved. We therefore recommend that further case studies should be conducted, covering a range of contrasting pesticides and scenarios. Consideration should be given to undertaking this as a cooperative research activity involving government, business, and academia. Risk managers as well as risk assessors and analysts should participate in the development of the case studies. In addition, efforts should be made to determine the practical significance of differences between competing theoretical approaches and to decide how these differences should be resolved for practical purposes.

2) Training in uncertainty analysis

There is a general need for increased training in uncertainty analysis, including a) basic training for all involved in the assessment and decision–making process; b) detailed training for risk assessors in using those methods that are adopted for routine use; and c) comprehensive training for risk analysts, so that they can use a broader range of methods and advise risk assessors and risk managers on their relative strengths and weaknesses.

3) Improvements in problem formulation

The use of uncertainty analysis and probabilistic methods requires systematic and detailed formulation of the assessment problem. To facilitate this, a) risk assessors and risk managers should be given training in problem formulation, b) tools to assist appropriate problem formulation should be developed, and c) efforts should be made to develop generic problem formulations (including assessment scenarios, conceptual models, and standard datasets), which can be used as a starting point for assessments of particular pesticides.

4) Standard distributions for pesticide assessments

There seems to be a desire among the workshop participants to develop a series of standard distributions, or distribution parameters, for exposure and effects variables that are generally used in risk assessments. In the case of toxicity data, for example, investigations leading to the quantification of a generic variance for between-species variation from pooled data for many pesticides may be useful (Luttik and Aldenberg 1997).

5) Improved methods for limited datasets

Limitations on data availability are a recurrent concern in discussions about uncertainty analysis and probabilistic methods, but arguably these methods are most needed when data are limited. More work is required to develop tools, methods, and guidance for dealing with limited datasets. Specific aspects that require attention are the treatment of sampling error in probability bounds analysis, and the use of qualitative information and expert judgment.

6) Improved methods and data for spatial and temporal variation

Spatial and temporal variation are major drivers of variability in risk. Research is required to develop the databases, models, and methods required to quantify their influence in risk assessment.

7) Evaluate the performance of probabilistic assessments

Methods for evaluating the performance and utility of uncertainty analysis in the context of probabilistic pesticide assessments are needed. This should include comparisons between assessment outputs and existing field data (e.g., avian field studies) to evaluate whether decision makers can rely on the assessment methods. Consideration should also be given to existing field data to refine generic assessment models, using Bayesian updating methods.

8) Improved processes for communication

Improved processes of communication among risk assessors, risk managers, and other stakeholders is needed.

9) Improved software

Efforts should be made to provide and improve user-friendly software, especially for those approaches where it currently appears to be lacking (e.g., Bayesian methods and Monte Carlo with more than 2 dimensions).

10) Improved access to resources for uncertainty analysis is needed

The internet should be used to maximize accessibility of software, databases, case study examples, and guidance and training materials.

11) International cooperation

International cooperation and, if possible, harmonization is desirable in developing new approaches, implementing the approaches, and using the approaches.

10.13 REFERENCES

Dakins ME, Toll JE, Small MJ, Brand KP. 1996. Risk-based environmental remediation: Bayesian Monte Carlo analysis and the expected value of sample information. Risk Anal 16 (1):67–79.

Dakins ME, Toll JE, Small W. 1994. Risk based environmental remediation: Decision framework and the role of uncertainty. Environ Toxicol Chem 13(12):1907–1915.

Luttik R, Aldenberg T. 1997. Extrapolation factors for small samples of pesticide toxicity data: Special focus on LD50 values for birds and mammals. Environ Toxicol Chem 16:1785–1788.

Glossary

2D Monte Carlo: A kind of nested Monte Carlo simulation in which distributions representing both incertitude and variability are combined together. Typically, the outer loop selects random values for the parameters specifying the distributions used in an inner loop to represent variability.

Accuracy: The degree of agreement between observed or measured values and the true value. Accuracy includes a combination of random error (precision) and systematic error (bias) components.

Aleatory uncertainty: The kind of uncertainty resulting from randomness or unpredictability due to stochasticity. Aleatory uncertainty is also known as variability, stochastic uncertainty, Type I or Type A uncertainty, irreducible uncertainty, conflict, and objective uncertainty.

Alpha error: See Type I error.

Arithmetic mean: A measure of central tendency. It is calculated as the sum of all the values of a set of measurements divided by the number of values in the set.

Assessment endpoint: An explicit expression of the environmental value that is to be protected, operationally defined by an ecological entity and its attributes. For example, salmon are valued ecological entities; reproduction and age-class structure are some of their important attributes. Together "salmon reproduction and age-class structure" form assessment endpoints.

Bayes' theorem: Original work by Sir Thomas Bayes, 1763. Composed of three pieces: 1) the prior distribution indicates the degree of belief about a random variable that exists before data are collected, 2) the likelihood function indicates the functional relationship of the data (experimental results) at the time of collection, and 3) the posterior distribution indicates the updated degree of believe. Thus, Bayes' theorem is a mathematical procedure for updating prior belief about a random variable, subsequent to observing new information. Bayes' theorem provides the underpinnings of decision-theoretic methods. Inferences drawn from Bayesian methods are fundamentally different than those derived from sampling theory, thus Bayes' theorem is a distinct paradigm for statistical inference and decision.

Beta error: See Type II error.

Bias: The systematic or persistent distortion of an estimate from the true value. From sampling theory, bias is a characteristic of the sample estimator of the sufficient statistics for the distribution of interest. Therefore, bias is not a function of the data, but of the method for estimating the population statistics. For example, the method for calculating the sample mean of a normal distribution is an unbiased estimator of the true but unknown population mean. Statistical bias is not a Bayesian concept, because Bayes' theorem does not relay on the long-term frequency expectations of sample estimators.

Bootstrap sample: A sample (e.g., 5000) obtained from an original data set by randomly drawing, with replacement, 5000 values from the original sample or a distribution estimated for that sample.

Bound: An upper bound of a set of real numbers is a real number that is greater than or equal to every number in the set. A lower bound is a number less than or equal to every number in the set. In this book, we also consider bounds on functions. These are not bounds on the range of the function, but rather bounds on the function for every function input. For instance, an upper bound on a function $F(x)$ is another function $B(x)$ such that $B(x) \geq F(x)$ for all values of x. $B(x)$ is a lower bound on the function if the inequality is reversed. If an upper bound cannot be any smaller, or a lower bound cannot be any larger, it is called a best possible bound.

Composition: The formation of one function by sequentially applying two or more functions. For example, the composite function $f(g(x))$ is obtained by applying the function g to the argument x and then applying the function f to this result.

Confidence interval: The numerical interval constructed around a point estimate of a population parameter. It is combined with a probability statement linking it to the populations' true parameter value, for example, a 90% confidence interval. If the same confidence interval construction technique and assumptions are used to calculate future intervals, they will include the unknown population parameter with the same specified probability. For example a 90% confidence interval around an arithmetic mean implies that 90% of the intervals calculated from repeated sampling of a population will include the unknown (true) arithmetic mean.

Conjugate pair: In Bayesian estimation, when the observation of new data changes only the parameters of the prior distribution and not its statistical shape (i.e., whether it is normal, beta, etc.), the prior distribution on the estimated parameter and the distribution of the quantity (from which observations are drawn) are said to form a conjugate pair. In case the likelihood and prior form a conjugate pair, the computational burden of Bayes' rule is greatly reduced.

Convolution: The mathematical operation that finds the distribution of a sum of random variables from the distributions of its addends. The term can be generalized to refer to differences, products, quotients, etc. It can also be generalized to refer to intervals, p-boxes and Dempster-Shafer structures as well as distributions.

Copula: The function that joins together marginal distributions to form a joint distribution function. For the bivariate case, a copula is a function C: $[0,1] \times [0,1] \circledR [0,1]$ such that $C(a, 0) = C(0, a) = 0$ for all $a \in [0,1]$, $C(a, 1) = C(1, a) = a$ for all $a \in [0,1]$, and $C(a_2, b_2) - C(a_1, b_2) - C(a_2, b_1) + C(a_1, b_1) \geq 0$ for all $a_1, a_2, b_1, b_2 \in [0,1]$ whenever $a_1 \leq a_2$ and $b_1 \leq b_2$. For any copula C, $\max(a + b - 1, 0) \leq C(a,b) \leq \min(a,b)$.

Credible interval: In a Bayesian analysis, the area under the posterior distribution. Represents the degree of belief, including all past and current information, of the random variable of interest. This term should not be confused with the confidence interval used in classical statistics.

Cumulative distribution function (CDF): The CDF is referred to as the "distribution function," "cumulative frequency function," or the "cumulative probability function." The cumulative distribution function, F(x), expresses the probability that a random variable X assumes a value less than or equal to some value x, F(x) = Prob (X > x). For continuous random variables, the cumulative distribution function is obtained from the probability density function by integration, or by summation in the case of discrete random variables.

Degree of belief: A Bayesian statistical concept that represents the state of information available to the investiagator concerning a random variable of interest. Belief is strengthened when past and current information are combined to give the investigator a good understanding of the random variable. Belief is measured as the area under the posterior distribution resulting from the implementation of Bayes' theorem.

Dempster-Shafer structure: A kind of uncertain number representing indistinguishability within bodies of evidence. In this book, a Dempster-Shafer structure is a finite set of closed intervals of the real line, each of which is associated with a nonnegative value m, such that the sum of all such m's is 1.

Deterministic methods: Methods in which all biological, chemical, physical, and environmental parameters are assumed to be constant and accurately specified.

Ecological risk assessment: The process that evaluates the likelihood that adverse ecological effects of differing magnitudes may occur or are occurring as a result of exposure to one or more stressors.

Epistemic uncertainty: The kind of uncertainty arising from imperfect knowledge. Epistemic uncertainty is also known as incertitude, ignorance, subjective uncertainty, Type II or Type B uncertainty, reducible uncertainty, nonspecificity and state-of-knowledge uncertainty.

Expert: A person who has 1) training and experience in the subject area resulting in superior knowledge in the field; 2) access to relevant information; 3) an ability to process and effectively use the information; and 4) is recognized by his or her peers or those conducting the study as qualified to provide judgments about assumptions, models, and model parameters at the level of detail required.

Expert judgment: A critical source of information based upon the collective experience of a scientist or expert in a particular field of study. For Bayesians, expert judgement is frequently used to form the prior distribution, thus formally incorporating an expert's degree of belief into statistical procedures.

Fuzzy arithmetic: Fuzzy arithmetic is the arithmetic embodied in operations such as addition, subtraction, multiplication, and division of fuzzy numbers. Fuzzy numbers are unimodal distribution functions of the real line that grade all real numbers according to the possibility that each might be a value the fuzzy number could take on. The minimum of the function is 0, which represents impossible values, and the maximum is 1, which represents those values that are entirely possible values that the fuzzy number might represent. Fuzzy numbers can be represented as a stack of intervals at each of infinitely many levels from 0 to 1, and fuzzy arithmetic can be thought of as a tool that permits propagation of the uncertainty represented by fuzzy numbers through mathematical models.

Hierarchical model: A model consisting of multiple parameters that can be regarded as related or connected in some way by the structure of the problem, implying that a joint probability model for these parameters should reflect the dependence among them.

Hypothesis testing: In classical statistics, a formal procedure for testing the long-term expected truth of a stated hypothesis. The statistical method involves comparison of two or more sets of sample data. On the basis of an expected distribution of the data, the test leads to a decision on whether to accept the null hypothesis (usually that there is no difference between the samples) or to reject that hypothesis and accept an alternative one (usually that there is some difference between the samples).

Imprecise probabilities: Any of several theories involving models of uncertainty that do not assume a unique underlying probability distribution, but instead correspond to a set of probability distributions. An imprecise probability arises when one's lower probability for an event is strictly smaller than one's upper probability for the same event. Theories of imprecise probabilities are often expressed in terms of a lower probability measure giving the lower probability for every possible event from some universal set, or in terms of closed convex sets of probability distributions (which are generally much more complicated structures than either probability boxes or Dempster-Shafer structures).

Incertitude: The kind of uncertainty arising from imperfect knowledge. Incertitude is also known as epistemic uncertainty, ignorance, subjective uncertainty, Type II or Type B uncertainty, reducible uncertainty, nonspecificity, and state-of-knowledge uncertainty.

Interval: A kind of uncertain number consisting of the set of all real numbers lying between two fixed numbers called the endpoints of the interval. In this book, intervals are always closed so that the endpoints are always considered part of the set.

Latin hypercube sampling (LHS): In Monte Carlo analysis, 1 of 2 sampling schemes are generally employed: simple random sampling or Latin hypercube sampling. Latin hypercube sampling may be viewed as a stratified sampling scheme designed to ensure that the upper or lower ends of the distributions used in the analysis are well represented. Latin hypercube sampling is considered to be more efficient than simple random sampling, that is, it requires fewer simulations to produce the same level of precision. Latin hypercube sampling is generally recommended over simple random sampling when the model is complex or when time and resource constraints are an issue.

Measurement endpoint: A measurable ecological characteristic that is related to the valued characteristic chosen as the assessment endpoint. Also known as "measure of effect."

Measurement error: The error inherent in the inability of a measuring device or procedure to provide an accurate representation of reality.

Median: The middle value for an ordered set of n values. It is represented by the central value when n is odd or by the mean of the two most central values when n is even.

Monte Carlo analysis: A modeling technique where parameter values are drawn at random from defined input probability distributions, combined according to the model equation, and the process repeated iteratively until a stable distribution of solutions results.

Nonparametric technique: A statistical technique that does not depend for its validity upon the assumption that the data were drawn from a specific distribution, such as the normal or lognormal. A distribution-free technique.

Parameter: Two distinct definitions for parameter are used. In the first usage (preferred), parameter refers to the constants characterizing the probability density function or cumulative distribution function of a random variable. For example, if the random variable W is known to be normally distributed with mean μ and standard deviation σ, the constants μ and σ are called parameters. In the second usage, parameter can be a constant or an independent variable in a mathematical equation or model. For example, in the equation $Z = X + 2Y$, the independent variables (X, Y) and the constant (2) are all parameters.

Population: In statistics and sampling design, the total universe addressed in a sampling effort.

Power: The probability of rejecting the null hypothesis in a statistical test when a particular alternative hypothesis happens to be true.

Precision: The degree to which a set of observations or measurements of the same property, usually obtained under similar conditions, conform to themselves.

Probability: The Bayesian or subjective view is that the probability of an event is the degree of belief that a person has, given some state of knowledge, that the event will occur. In the classical or frequentist view, the probability of an event is the frequency of an event occurring given a long sequence of identical and independent trials.

Probability box: A kind of uncertain number representing both incertitude and variability. A probability box can be specified by a pair of functions serving as bounds about an imprecisely known cumulative distribution function. The probability box is identified with the class of distribution functions that would be consistent with (i.e., bounded by) these distributions.

Probability density function (PDF): The PDF is referred to as the "probability function" or the "frequency function." For continuous random variables, that is, the random variables that can assume any value within some defined range (either finite or infinite), the probability density function expresses the probability that the random variable falls within some very small interval. For discrete random variables, that is, random variables that can only assume certain isolated or fixed values, the term "probability mass function" (PMF) is preferred over the term "probability density function." PMF expresses the probability that the random variable takes on a specific value.

Propagation of error: Mathematical technique for computing the total error of a model prediction by calculating the error for each term in the model, and propagating the errors through the model into the total error of prediction.

Quantile: The value in a distribution that corresponds to a specified proportion of the population distribution or distribution function. Quartiles (25th, 50th, and 75th percentiles), the median (50th percentile), and other percentiles are special cases of quantiles.

Random: A chance event.

Risk: Two definitions are commonly used. The 1st states that risk is the probability of a prescribed undesired effect. The 2nd states that risk is the relationship between probability and magnitude of effect.

Risk management: A decision-making process that considers political, social, economic, and technological information in conjunction with risk assessment information to select an appropriate response to a particular problem.

Robust Bayes: A school of thought among Bayesian analysts in which epistemic uncertainty about prior distributions or likelihood functions is quantified and projected through Bayes rule to obtain a class of posterior distributions.

Safety factor: A factor applied to an observed or estimated toxic concentration or dose to arrive at a safe level.

Sampling error: In surveys, investigators frequently take measurements (or samples) on the parameters of interest, from which inferences to the true but unknown population are inferred. The inability of the sample statistics to represent the true population statistics is called sample error. There are many reasons why the sample may be inaccurate, from the design of the experiment to the inability of the measuring device. In some cases, the sources of error may be separated (see Variance components).

Screening-level risk assessment: A risk assessment methodology that identifies stressors of potential concern and eliminates from further consideration those not posing any significant risk.

Sensitivity, sensitivity analysis: Mathematical technique for determining the relative influence of an individual model parameter(s) on the predicted value. A related term, elasticity, is defined as the relative change in model prediction over the relative change in the parameter value. In a broader sense, sensitivity can refer to how conclusions change if models, data, or assessment assumptions are changed.

Standard deviation: A common measure of the dispersion or imprecision of observed values expressed as the positive square root of the variance.

Statistic: A computed or estimated quantity such as the mean, standard deviation, or correlation coefficient.

Stochastic: A process involving a random variable.

Type I error (alpha error): An incorrect decision resulting from rejecting the null hypothesis when the null hypothesis is true. A false positive decision.

Type II error (beta error): An incorrect decision resulting from failing to reject the null hypothesis when the alternative hypothesis is true. A false negative decision.

Uncertainty: Imperfect knowledge concerning the present or future state of the system under consideration; a component of risk resulting from imperfect knowledge of many kinds including the degree of hazard or of its spatial and temporal pattern of expression. Generally, uncertainty can be reduced with further information and knowledge.

Uncertainty analysis: Determination of the sources of uncertainty in the measurement or prediction of environmental parameters. The analysis can be both quantitative (computation of variances) or qualitative (lists of uncertain methods and procedures). The total uncertainty in the parameters of interest is typically a function of all of the individual sources of uncertainty.

Uncertainty factor: A factor applied to an exposure or effects concentration or dose to correct for identified sources of uncertainty.

Variability: Variability refers to observed differences attributable to true heterogeneity or diversity in a population or exposure parameter. Variability is the result of natural random processes and stems from environmental, lifestyle, and genetic differences. Examples include physiological variation (e.g., natural variation in body weight, height, breathing rates, drinking water intake rates), weather variabilty, variation in soil types, and differences in contaminant concentrations in the environment. From statistical sampling theory, the true variability is fixed, but the sample estimate of the population variance can be reduced by further measurement or study.

Variance: A measure of the dispersion of a set of values. It is calculated by taking the difference between each individual value of a set and the arithmetic mean of the set, squaring each difference, summing the squares, and then dividing the sum by one less than the number of values in the set.

Variance components: A statistical technique for factoring the total variance in a random parameter into its component parts. Typically, a model is defined that represents the experimenter's understanding of the variance components. This model is used to separate the variance components. The model is called a variance components model.

Weight of evidence: The result of an evaluation of multiple lines of evidence in an ecological risk assessment. A weight of evidence approach reduces many of the biases and uncertainties associated with using only one approach to estimate risk. The lines of evidence that may be considered in a weight of evidence approach include comparing levels in the environment to the results of laboratory bioassays, field observations, in situ tests, ecoepidemiology, and population and ecosystem modeling. Each line of evidence is evaluated for relevance of the evidence to the exposure scenario of interest, relevance of the evidence to the assessment endpoint, confidence in the evidence or risk estimate, and likelihood of causality.

Index

Other Titles from the Society of Environmental Toxicology and Chemistry (SETAC)

Genomics in Regulatory Ecotoxicology: Applications and Challenges
Ankley, Miracle, Perkins, Daston, editors
2007

Population-Level Ecological Risk Assessment
Barnthouse, Munns, Sorensen, editors
2007

*Effects of Water Chemistry on Bioavailability and Toxicity of Waterborne Cadmium,
Copper, Nickel, Lead, and Zinc on Freshwater Organisms*
Meyer, Clearwater, Doser, Rogaczewski, Hansen
2007

Ecosystem Responses to Mercury Contamination: Indicators of Change
Harris, Krabbenhoft, Mason, Murray, Reash, Saltman, editors
2007

Freshwater Bivalve Ecotoxicology
Farris, Van Hassel, editors
2006

*Estrogens and Xenoestrogens in the Aquatic Environment:
An Integrated Approach for Field Monitoring and Effect Assessment*
Vethaak, Schrap, de Voogt, editors
2006

*Assessing the Hazard of Metals and Inorganic Metal Substances
in Aquatic and Terrestrial Systems*
Adams, Chapman, editors
2006

Perchlorate Ecotoxicology
Kendall, Smith, editors
2006

Natural Attenuation of Trace Element Availability in Soils
Hamon, McLaughlin, Stevens, editors
2006

*Mercury Cycling in a Wetland-Dominated Ecosystem:
A Multidisciplinary Study*
O'Driscoll, Rencz, Lean
2005

Atrazine in North American Surface Waters:
A Probabilistic Aquatic Ecological Risk Assessment
Giddings, editor
2005

Effects of Pesticides in the Field
Liess, Brown, Dohmen, Duquesne, Hart, Heimbach, Kreuger, Lagadic,
Maund, Reinert, Streloke, Tarazona
2005

Human Pharmaceuticals: Assessing the Impacts on Aquatic Ecosystems
Williams, editor
2005

Toxicity of Dietborne Metals to Aquatic Organisms
Meyer, Adams, Brix, Luoma, Stubblefield, Wood, editors
2005

Toxicity Reduction and Toxicity Identification Evaluations for Effluents, Ambient
Waters, and Other Aqueous Media
Norberg-King, Ausley, Burton, Goodfellow, Miller, Waller, editors
2005

Use of Sediment Quality Guidelines and Related Tools
for the Assessment of Contaminated Sediments
Wenning, Batley, Ingersoll, Moore, editors
2005

Life-Cycle Assessment of Metals
Dubreuil, editor
2005

Working Environment in Life-Cycle Assessment
Poulsen, Jensen, editors
2005

Life-Cycle Management
Hunkeler, Saur, Rebitzer, Finkbeiner, Schmidt, Jensen, Stranddorf, Christiansen
2004

Scenarios in Life-Cycle Assessment
Rebitzer, Ekvall, editors

Ecological Assessment of Aquatic Resources:
Linking Science to Decision-Making
Barbour, Norton, Preston, Thornton, editors
2004

Life-Cycle Assessment and SETAC: 1991–1999
15 LCA publications on CD-ROM
2003

SETAC

A Professional Society for Environmental Scientists and Engineers and Related Disciplines Concerned with Environmental Quality

The Society of Environmental Toxicology and Chemistry (SETAC), with offices currently in North America and Europe, is a nonprofit, professional society established to provide a forum for individuals and institutions engaged in the study of environmental problems, management and regulation of natural resources, education, research and development, and manufacturing and distribution.

Specific goals of the society are

- Promote research, education, and training in the environmental sciences.
- Promote the systematic application of all relevant scientific disciplines to the evaluation of chemical hazards.
- Participate in the scientific interpretation of issues concerned with hazard assessment and risk analysis.
- Support the development of ecologically acceptable practices and principles.
- Provide a forum (meetings and publications) for communication among professionals in government, business, academia, and other segments of society involved in the use, protection, and management of our environment.

These goals are pursued through the conduct of numerous activities, which include:

- Hold annual meetings with study and workshop sessions, platform and poster papers, and achievement and merit awards.
- Sponsor a monthly scientific journal, a newsletter, and special technical publications.
- Provide funds for education and training through the SETAC Scholarship/Fellowship Program.
- Organize and sponsor chapters to provide a forum for the presentation of scientific data and for the interchange and study of information about local concerns.
- Provide advice and counsel to technical and nontechnical persons through a number of standing and ad hoc committees.

SETAC membership currently is composed of more than 5000 individuals from government, academia, business, and public-interest groups with technical backgrounds in chemistry, toxicology, biology, ecology, atmospheric sciences, health sciences, earth sciences, and engineering.

If you have training in these or related disciplines and are engaged in the study, use, or management of environmental resources, SETAC can fulfill your professional affiliation needs.

All members receive a newsletter highlighting environmental topics and SETAC activities and reduced fees for the Annual Meeting and SETAC special publications.

All members except Students and Senior Active Members receive monthly issues of Environmental Toxicology and Chemistry (ET&C) and Integrated Environmental Assessment and Management (IEAM), peer-reviewed journals of the Society. Student and Senior Active Members may subscribe to the journal. Members may hold office and, with the Emeritus Members, constitute the voting membership.

If you desire further information, contact the appropriate SETAC Office.

1010 North 12th Avenue
Pensacola, Florida 32501-3367 USA
T 850 469 1500   F 850 469 9778
E setac@setac.org

Avenue de la Toison d'Or 67
B-1060 Brussels, Belgium
T 32 2 772 72 81   F 32 2 770 53 86
E setac@setaceu.org

www.setac.org
Environmental Quality Through Science®